Reciprocal Frame Architecture

To Jens and Sofia

RECIPROCAL FRAME ARCHITECTURE

Olga Popovic Larsen

AMSTERDAM • BOSTON • HEIDELBERG • LONDON • NEW YORK • OXFORD
PARIS • SAN DIEGO • SAN FRANCISCO • SINGAPORE • SYDNEY • TOKYO
Architectural Press is an imprint of Elsevier

Architectural Press is an imprint of Elsevier
Linacre House, Jordan Hill, Oxford OX2 8DP, UK
30 Corporate Drive, Suite 400, Burlington, MA 01803, USA

First published 2008

British Library Cataloguing-in-Publication Data
A catalogue record for this title is available from the British Library

Library of Congress Cataloging-in-Publication Data
A catalog record for this title is available from the Library of Congress

ISBN: 978-0-7506-8263-3

For information on all Architectural Press publications visit our website at www.architecturalpress.com

Typeset by Charon Tec Ltd (A Macmillan Company), Chennai, India
www.charontec.com

Printed and bound in Slovenia

08 09 10 11 12 10 9 8 7 6 5 4 3 2 1

CONTENTS

Foreword

This book covers the little known structural and architectural concept, design and construction of reciprocal frames, and is the first authoritive book of its kind, with an exhaustive coverage of a multitude of types.

A simple description of reciprocal frames is 'a structure made up of mutually supporting beams in a closed circuit' – quite a good definition without a diagram or model. I have a six-membered timber model, made by Dr Popovic, which beautifully illustrates the simple principles.

History has many examples – Serlio, da Vinci and Villard de Honnecourt – but these early ones were all planar examples. Here a huge variety of types are analysed and illustrated.

This is a specialist's book, with perhaps a limited appeal to architects and engineers at the forefront of thinking, but is fascinating as a treatise on an unusual structural system. Its content and scope are incredibly comprehensive, particularly on its extensive coverage of the many buildings in Japan, where the majority of the research was carried out. A 'mind blowing' book, which I am sure will lead to more exploration of 'reciprocal frame structures' in the future.

Professor Tony Hunt

Acknowledgements

I would like to express my deep gratitude to a number of people who have helped in different ways to bring this book to completion.

The participation and enthusiasm of the designers, Kazuhiro Ishii, Yoichi Kan, Tadashi Hamauzu, Graham Brown and John Chilton, whose projects are discussed in the case studies was both a vital factor and an inspiration. I am especially grateful to the Japanese designers, who spent many hours talking to me about their designs, which helped me understand the philosophical depth of their work. In that regard I would like to mention architect Hiroshi Sawazaki, Managing Director (President) of Keikaku-Inc., who kindly agreed to talk to me about the work of his deceased colleague, architect Yasufumi Kijima, one of the founders of Keikaku-Inc., whose Stonemason Museum is presented in the book. The travelling in southern Japan was organized by Mr Yoichi Kan, Managing Director (President) of Pal Corporation and one of the RF designers featured, with his design of the New Farmhouse reciprocal frame building. Mrs Keiko Miyahara was great company, and I am grateful to her for helping me understand the subtleties of the refined Japanese culture.

I am grateful to researchers John Chilton, Olivier Baverel, Masseoud Saidani, Joe Rizzuto and Vito Bertin, who kindly provided up-to-date information about their work.

Designers Tony Wrench, Hugh Adamson and Fred Oesch helped in providing information about their recent projects using the reciprocal frame structure.

The assistance of architect Chris Dunn of Whitbybird, who helped me with the parametric studies, is greatly appreciated.

Structural engineer Jens Larsen of Ove Arup Sheffield helped with the modelling and structural analysis of reciprocal frames.

The marvellous hand-redrawn images are the work of Amir Ebrahim Piroozfar (Poorang), architect and Ph.D. candidate at the University of Sheffield School of Architecture. Poorang spent a great deal of his own time trying to convert my suggestions into meaningful images. His assistance with scanning, preparing images, converting files and collating the material for the book at a particularly busy time of year is gratefully acknowledged.

(All uncredited photographs and sketches are those of the author.)

The translation work done by Damien Osaka, who translated from Japanese the writings about architects Ishii and Kijima, was a great help in understanding the Japanese texts.

I have dealt with several people at Architectural Press, including Alex Hollingsworth and Jodi Cusack, both of whom have been supportive in helping me produce the final text.

The trip to Japan was funded by the Great Britain Sasakawa Foundation and by Elsevier, which helped enormously.

For one semester I had the opportunity to work on the research for my book. I am grateful to the School of Architecture, University of Sheffield, for granting me leave from my everyday teaching duties. This enabled me to travel to Japan, meet the Japanese reciprocal frame designers and understand better the masterpieces of reciprocal frame architecture. Without this leave the already tight deadline for producing the manuscript would not have been possible.

I am most grateful to Dr Colin Roth for reading my English and for his comments on how to improve it.

My recently deceased friend Di Ramsamy helped in many ways. She was a great listener and while writing the book Di helped by giving me both moral and practical support. I am also very grateful to her for helping out with child care.

Without naming them all, I would like to say a big thank you to friends, colleagues and family for their support and help throughout the writing process. Finally, a thank you to Jens for his continued encouragement and advice on the text of the book. Last, but not least: huge thanks to Sofia for taking it so well that '*mummy has to work late again*!'

1 Introduction

The title of this book, *Reciprocal Frame Architecture*, asserts that this is a book about architecture, but why 'reciprocal frame' architecture? What are 'reciprocal frames'? The term means hardly anything, even to people who are in the field, like architects and engineers (unless they are already familiar with it for one reason or another). To ordinary people the name 'reciprocal frame' certainly does not mean much. This is perhaps one of the reasons for writing this book – to make reciprocal frame structures and the architectural forms they create better known.

Before talking about the opportunities that reciprocal frames offer, one has to start by defining the meaning of the title. From the name one can easily get the impression that the subject belongs to the field of frames, but then why 'reciprocal'? Frames are a well-established structural system. What does 'reciprocal' signify when describing a structure and what kind of quality does it add to frame structure, if any at all? Also, what is the connection to architecture? What is 'reciprocal frame architecture'?

We will start by defining the meaning of the terms used in the title, 'reciprocal frame' and 'architecture', and establish what they signify.

The reciprocal frame is a three-dimensional grillage structure mainly used as a roof structure, consisting of mutually supporting sloping beams placed in a closed circuit. The inner end of each beam rests on and is supported by the adjacent beam. At the outer end the beams are supported by an external wall, ring beam or by columns. The mutually supporting radiating beams placed tangentially around a central point of symmetry form an inner polygon. The outer ends of the beams form an outer polygon or a circle. If the reciprocal frame (RF) is used as a roof structure, the inner polygon gives an opportunity of creating a roof light.

The RF principle is not new and has been used throughout history, especially in the form of a flat configuration. This variation of the RF, where the beams are connected in the same plane forming a planar grillage, is presented in detail in Chapter 2. Flat grillages have typically been used for forming ceilings and floors when timbers of sufficient length were not available. Examples are the structures developed by Serlio, da Vinci, Honnecourt and others presented in Chapter 2. None

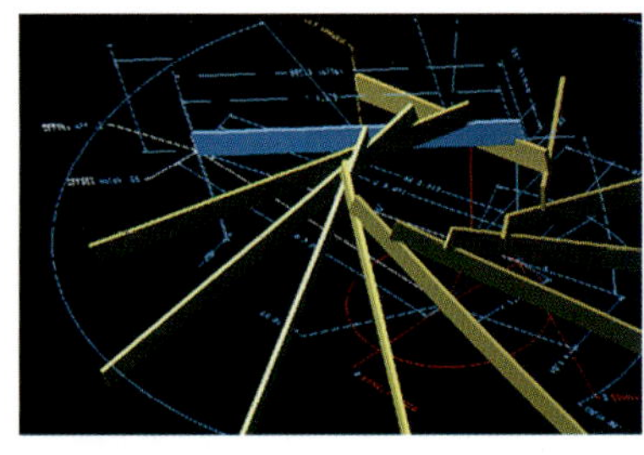

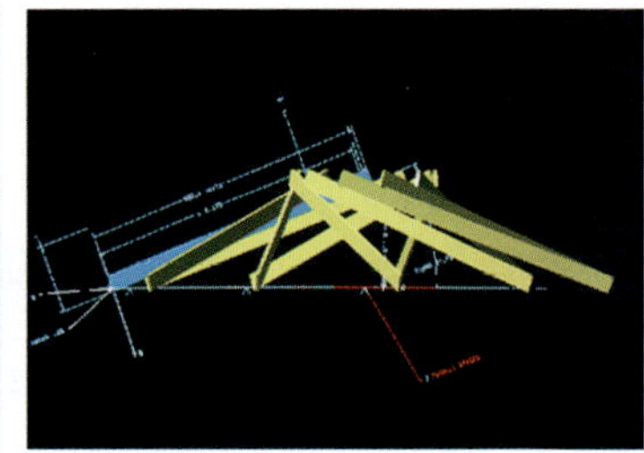

▲ **1.1** Typical RF structure - 3D view, elevation and detail.

of these designers, however, used the name 'reciprocal frame' for their structures.

The name 'reciprocal frame' comes from Graham Brown, who developed this type of structure in the UK. Graham used 'reciprocal' because of the way the beams mutually support each other.

In the Oxford English Dictionary the word 'reciprocal' has several meanings:

- Mathematical – so related to another that their product is unity
- Adjective – in return (for example, I helped him and he helped me in return).

In our context, it represents the appearance and behaviour of the unified structure in which each beam supports, and in turn is supported by, all of the others.

Because of the geometrical characteristics of the structure, the most appropriate forms of buildings (in plan) using the RF are circular, elliptical and regular polygonal. As a result, so far most of the buildings constructed using the RF have regular polygonal or circular plans. In the case of regular plan forms, all RF members are identical, which gives the possibility of modular RF construction.

The circular plan form was one of the first used. Many vernacular buildings throughout human history (mud huts, cave dwellings and so on) had approximately circular plan forms. They would appear to have a protective, womb-like quality. Also, circular and regular polygonal forms are typical in buildings of major significance, such as churches, concert halls, sports stadia, museums and the like.

If suitable materials are used for the main RF members, such as reinforced concrete, glued laminated timber, steel beams or trusses, the RF could span short and long distances with equal success. Because of the most common plan forms, polygonal and circular, the organization of the function and division of the internal spaces of the RF buildings are different from buildings with rectilinear plan forms. Since no internal supports are needed, the RF forms a very flexible architectural space. It is important to note that the beams that form the RF do not meet in a central point (as shown in Figure 1.1). This is different to most of the roof structures over buildings with circular plan forms, which have radial members meeting at the highest point of the roof.

On the other hand, since the inner and the outer polygons are defined by the end points of the beams, which can have different lengths, the RF can be used to cover almost any form in plan. The possibility of creating an infinite variety of plan forms, and at the same time incorporating different spans, makes it possible for the structure to be used on buildings with very different functions – indeed, for any function. Because the structure is not very well known, and despite its great potential, not many buildings using RFs have been constructed to date.

If one looks at the structures designed by Pier Luigi Nervi, the elegant shells designed by Heinz Isler or the great biomes of the Eden Project, it is evident that structural form defines architectural form to a great extent. The RF, although very different in scale to the mentioned structures, is similar in that it also influences architectural forms. The visual impact of the structure of self-supporting spiralling beams is very powerful. It clearly not only makes the buildings stand up, but affects how the spaces can be used as well as the overall architectural expression.

By varying the geometrical parameters of the RF structure, such as the length and number of beams, radii of inner and outer polygons and the beam slopes, a designer can achieve a great number of variations of the same structural system. In addition, one has the option of using single or multiple RF units (a combination of several single units), which adds to the versatility of the system and creates different architectural expressions.

Like any structural form, the RF structure has its limitations. There is no such thing as 'the perfect structural solution' and this book is not trying to present the RF as such. Rather, it will present the opportunities the RF offers, but also describe the most common challenges that arise.

The RF is still relatively unknown to most professionals and its architectural potential remains largely unexplored. This book therefore aims to bring the RF closer to designers, clients and users, making it a viable option in building design.

This book is structured in two parts. The first part (Chapters 1–5) looks at historical precedents, investigating possible morphologies (forms) that can be created with RFs, defining the geometrical parameters of the structure and its structural behaviour. The second part of the book presents the work of Japanese RF designers Kazuhiro Ishii, Yoichi Kan and Yasufumi Kijima, as well as British designer Graham Brown. Chapter 6 shows the context in which the Japanese RF buildings have emerged, while the case studies of reciprocal frame architecture (Chapters 7–10) show examples in which the RF structure and the architecture complement each other to form 'reciprocal frame architecture'. Chapter 11 shows some additional recently built examples using RF structures.

Reciprocal frame architecture encompasses the work of many researchers and practitioners who have pushed the boundaries of what is possible in this field. The research and design work of John Chilton, pioneer in exploring the structural behaviour, geometry and morphology of RFs, is a valuable contribution. In addition, I also refer to the work of researchers Olivier Baverel, Messaoud Saidani, Joe Rizzuto, Vito Bertin and others, who have contributed to a better understanding of how these structures are configured and how they behave structurally.

The architectural work of designers Ishii, Kan, Kijima, Brown, Wrench, Adamson, Oesch and others shows what is possible in practice. Some of these designers who have contributed with their designs to reciprocal frame architecture have been able to demonstrate a real synthesis of structure and architecture, creating genuine architectural masterpieces.

It is hoped that this book will inspire the reader to learn more about the world of the reciprocal frame and how to use this amazing structure in creating new forms of architecture – reciprocal frame architecture.

2 Background – The Reciprocal Frame Historically

So who made the first reciprocal frame? Where did the idea come from? It would be difficult to find out when and where the first reciprocal frame (RF) was constructed; to do so would be like trying to establish when and where the first high-heeled shoe was produced, or when the first green wooden toy car was made. Perhaps these two would be easier to establish than the whereabouts of the first RF structures. There are two main reasons for this: the first is that very few people describe these structures as *reciprocal frames*; the second is that the idea is very old and the historic structures that adopted RF principles were mainly built of timber (well before steel and concrete were known to humankind), which deteriorated over the centuries or were lost in fires. Finding written documentation is not easy either, because of the absence of a common name for them.

Still, despite these difficulties which prevent us establishing where the first ideas about using structures like the RF originated, we can easily demonstrate that the RF principle has been around for many centuries.

Structures such as the neolithic pit dwelling (Figure 2.1), the Eskimo tent, Indian tepee (Figure 2.2) or the Hogan dwellings (Figure 2.3) have some similarities to the RF concept. Perhaps the latter two examples have greater similarities to the RF than the neolithic pit dwelling and the Eskimo tent. Similarly to the RF, the Indian tepee and the Hogan dwellings use the principle of mutually supporting beams. The differences between them and the RF are that the rafters forming the structure of the Indian tepee come together into a point where they are tied together and the integrity of the structure is secured in that way. Stretched animal skins provide additional stiffness to the conical form of the tepee. The animal skins have the role of the cladding roof panels used in RF structures, which in a similar way provide a 'stretched skin effect' and give additional stiffness to the structure.

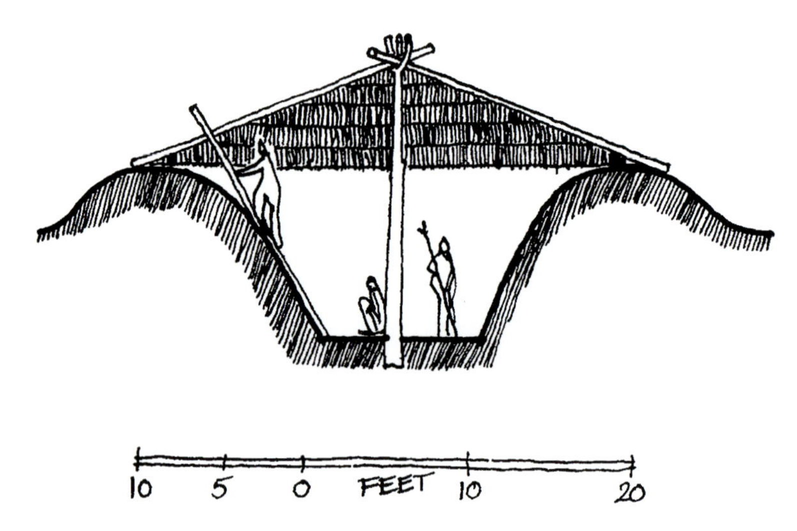

▲ **2.1** Neolithic pit dwelling. (Sketch by A. E. Piroozfar.)

▲ **2.2** Indian tepee. (Sketch by A. E. Piroozfar.)

▲ **2.3** Hogan dwelling. (Sketch by A. E. Piroozfar.)

The Hogan dwelling looks, in plan, very much like a complex RF structure consisting of a large number of single RFs being supported by a larger diameter RF structure, which in turn is inserted into and supported by an even larger RF. This configuration of a semi-regular form of the Hogan timber structure forms a domed roof. In most cases the Hogans are covered with mud, which not only provides a better internal climate, but also 'glues' the timber rafters together and creates a stable structural form.

Greater similarity to RFs can be seen in the later development of structural forms such as medieval floor grillages, Honnecourt's planar floor grillages, Leonardo da Vinci's structural sketches, as well as Sebastiano Serlio's and Wallis's RF-like structures.

As stated earlier, it is very difficult (if not impossible) to establish where the first RF structure was built. It is very likely that more than one civilization used structures similar to RFs. However, the only written data about structures similar to the present form of RFs can be found in Japan. There is evidence (Ishii, 1992/3) that in the late twelfth century the Buddhist monk Chogen (1121–1206) established a technique of spiral layering of wood beams which was used in construction of temples and shrines. Unfortunately, no buildings remain that have been constructed in this way. The timber structures have been destroyed by fires, wars or lost due to decay. It is important, though, to stress that the technique which Chogen used is identical to the structural principle of the RF, and it has been used as a roof structure on other, more recent buildings in Japan. These will be presented in detail through the case studies of Japanese contemporary RF buildings later in this book.

The geometric forms of these temples in plan are reminiscent of the mandalas used in Buddhist meditation as symbols of divinities, thus the name 'mandala dach' (mandala roof) has been used for the RF in Germany. 'Mandala' is a Sanskrit word meaning 'magic circle' (Gombrich, 1979) and it is a geometric pattern which includes circles and squares arranged to have symbolic significance. They are one of the oldest religious symbols, and can be found as painted decoration on ceilings in religious buildings such as Tun-huang in China.

The role of the mandala in meditation is described by Auboyer (1967, p. 26) in the following manner: 'The one who meditates on a mandala must "realize" through meditative effort and prayer the divinities belonging to each zone. Progress is toward the centre, at which point the person meditating attains mystical union with divinity.' On studying the form of the RF, it can be noted that the beams of the structure focus towards the central polygon which frames the sky or heaven to echo the role of the mandala. Some examples of mandalas are presented in Figure 2.4.

If we look at the history of Western architecture, it is evident that in medieval times most buildings were constructed with timber floors. The smaller buildings (such as houses and farm buildings) were built mainly in timber, whereas the more important buildings (such as churches or palaces) were built in stone (walls), with timber floors used to span between the walls and create the different levels in the building. As the

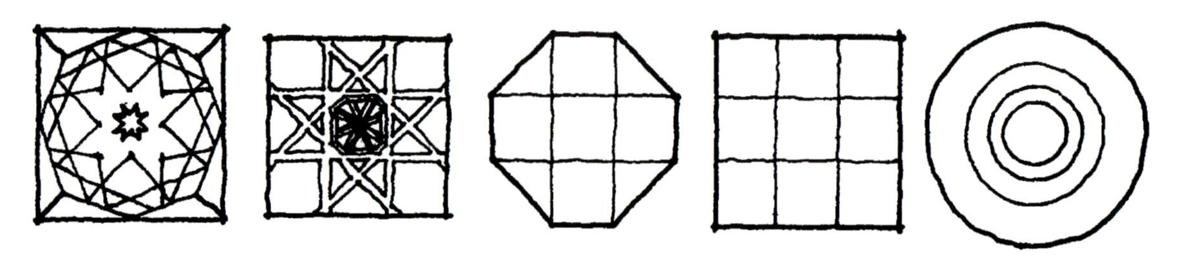

▲ 2.4 Mandala geometry. (Sketch by A. E. Piroozfar.)

▲ 2.5 Typical medieval floor grillage configuration. (Sketch by A. E. Piroozfar.)

buildings became bigger and had larger rooms, there was a need for timber that could span greater distances. Often, these great timbers had to be brought from far away but when this was not possible alternative floor designs were investigated. It is likely that in such circumstances a solution for spanning distances longer than the available beams was devised in the form of a beam grillage. Medieval floors were sometimes supported on four beams, all shorter than the span. This was also a common configuration for the framing of stairwells, as shown in Figure 2.5 (Chilton, Choo and Yu, 1994). These structures were usually planar grillages, but examples of three-dimensional structures can also be found. It is interesting that this 'medieval grillage structure' works in a similar way to the RF. It is actually a flat version of an RF with inner connections that transfer moments, as explained in more detail in the section of this book dedicated to structural behaviour (see Chapter 5).

One such medieval architect, Villard de Honnecourt, who studied the construction of great churches such as Cambrai, Rheims and Laon, and may even have been in charge of their building, provides us with information on how to deal with the problem of beams shorter than the span, or as he puts it: 'How to work on a house or tower even if the timbers are too short' (Bowie, 1959, p. 130).

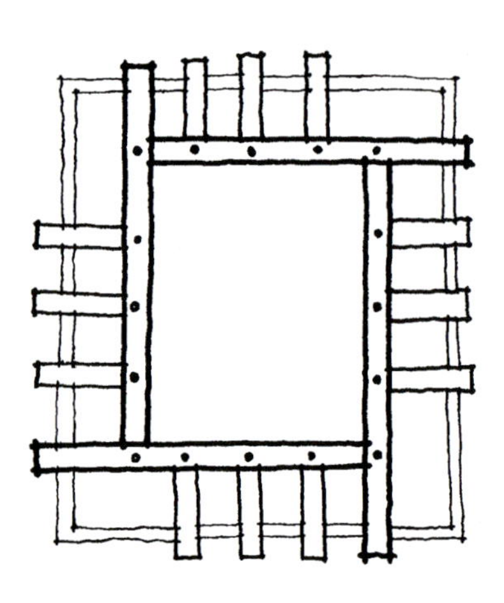

▲ 2.6 Honnecourt's planar grillage assembly. (Sketch by A. E. Piroozfar.)

Honnecourt gives no information on the spans he had in mind or where this solution has been applied, but some other authors do. Honnecourt's solution to this problem (presented in Figure 2.6) is a planar grillage and it adopts similar principles to the RF. If four beams in an RF were arranged so that they have no slope, and, instead of being placed on top of each other, if they were arranged and connected in the same plane, we would get Honnecourt's configuration. The difference is that an RF (with inclined members) transfers loads through compression in each member, whereas the flat configurations do not.

Honnecourt's sketches were made in the period 1225–1250. This indicates that these types of structure have been known for a very long time.

Although a great deal of research has been done on cathedral architecture, there is very little data on functional carpentry. This is perhaps because, as Hewett (1974, p. 9) stated, '. . . the roofs were normally hidden above stone vaults and only accessible with difficulty in darkness and dirt.'

There is evidence that flat configurations of structures similar to the RF have been used for polygonal chapter house roofing. An example of this is the chapter house at Lincoln, designed by Alexander and built in the period 1220–1235. The roof, which is of a puzzling complexity, encloses the ten-sided regular polygonal plan of the chapter house. 'It is a real master work, which comprises of two parts – the lower a "gambrel"-type decagonal structure, and the higher part, which restored the roof to a fully pyramidal form . . .' (Hewett, 1974, p. 74), as presented in Figures 2.7 and 2.8.

They are actually two superimposed queen-post assemblies set inside a pitched roof with a king post. The RF-like structure is at the base of the

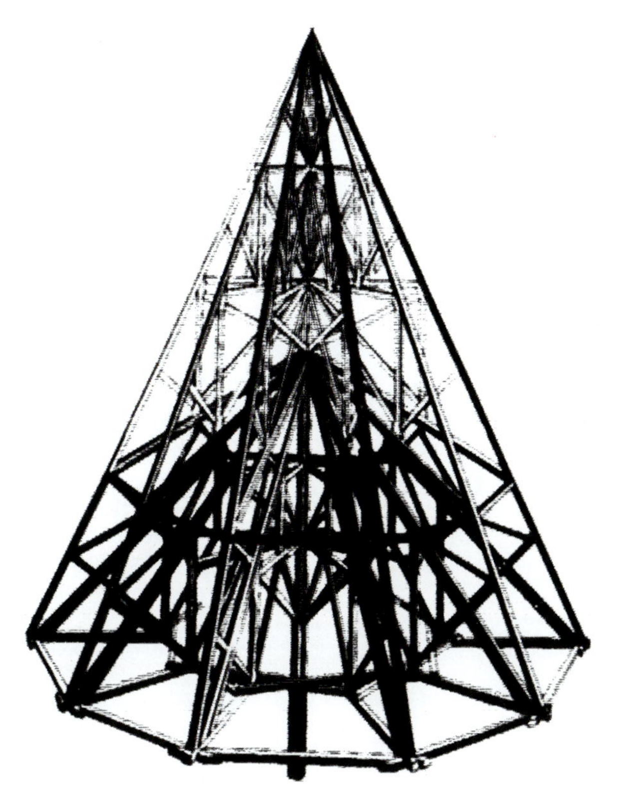

▲ **2.7** Roof of the chapter house at Lincoln cathedral – 3D view. (Sketch by A. E. Piroozfar.)

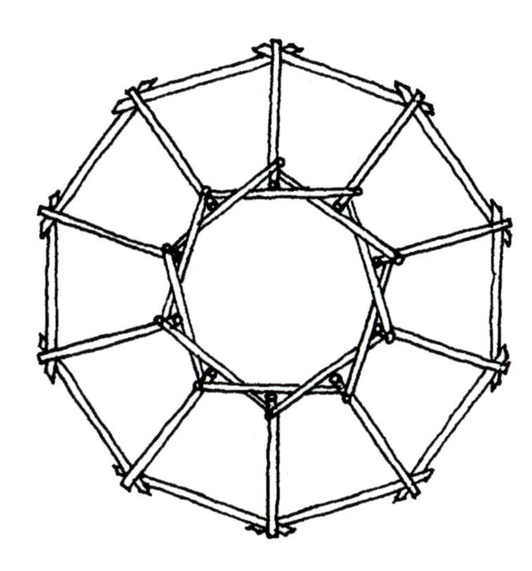

▲ **2.8** Roof of the chapter house at Lincoln cathedral – plan view. (Sketch by A. E. Piroozfar.)

roof, which was built of softwood (pine) and mainly held together by ironwork and forelock-bolts. It would have been better for the radial extension and shearing stresses to which the structure is subjected if it had been constructed from timber of higher quality, but it seems that cost was the reason behind the choice. This part of the roof structure is actually identical to a flat RF, and was probably used for the first time in roofs for polygonal spaces. Hewett describes it as 'ingenious' and says that '. . . the construction of the essential "ring-beam" secures the inner ends of the ten radiating ties and it is possibly the architect's invention' (Hewett, 1974, p. 81). Figure 2.8 shows the plan of this structure.

▲ **2.9** Flat beam grillage by Leonardo da Vinci. (Sketch by A. E. Piroozfar.)

Two hundred years later, Leonardo da Vinci (1452–1519), known as one of the greatest of Renaissance thinkers, who conducted studies in physics, anatomy, medicine, astronomy, fortification, canal-making, architecture and engineering, was also interested in structures very similar to the RF (Richter, 1977). His sketch in Volume I of the Codex Madrid (Figure 2.9) shows a planar grillage of four beams, identical to the main grillage structure proposed by Honnecourt (Figure 2.6). Leonardo also explored assemblies of beam grillages, which are presented in his sketches of the Codex Atlantico, as shown in Figures 2.10a and b.

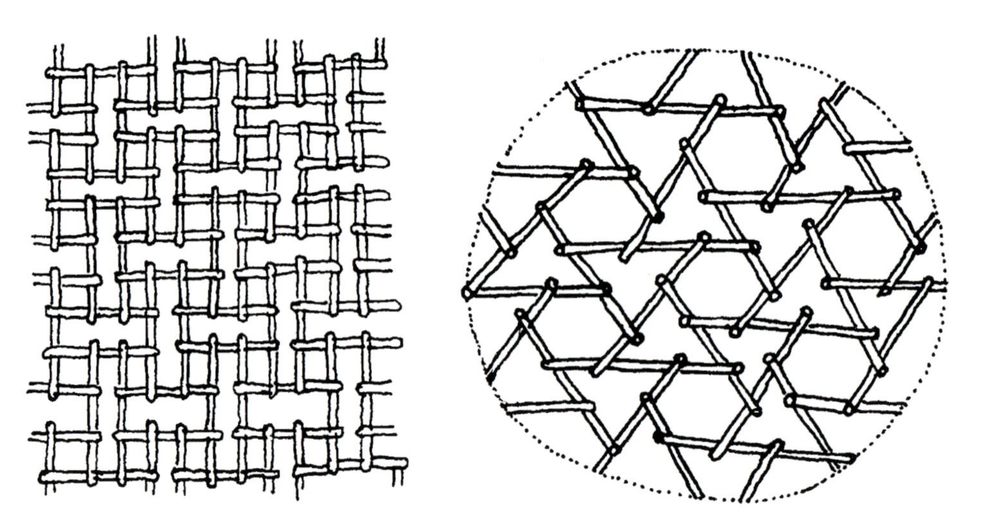

▲ **2.10** (a) and (b) Sketches of grillage assemblies by Leonardo da Vinci. (Sketches by A. E. Piroozfar.)

Leonardo da Vinci also made drawings of arched forms created by using short timbers for his bridge designs. Examples of these are the 'temporary bridges' (Anon, 1956), originally presented in Codex Atlantico (Figure 2.11a, b). They are constructed from relatively short timber beams which support and are being supported by each other. The three-dimensional structure is actually formed of two mutually connected two-dimensional arches built from the short timber beams. These types of bridges are known to be used in Chinese traditional architecture. A similar contemporary example is the bamboo pedestrian bridge in Rio de Janeiro, presented in Figure 2.12.

Leonardo's arched beams are very similar to the ring beam at the chapter house of Lincoln cathedral. The only difference is that the latter is a whole circle ring beam, whereas Leonardo's bridges are created by beams that form a segmented arch. Both structures, to some degree, are similar to an RF.

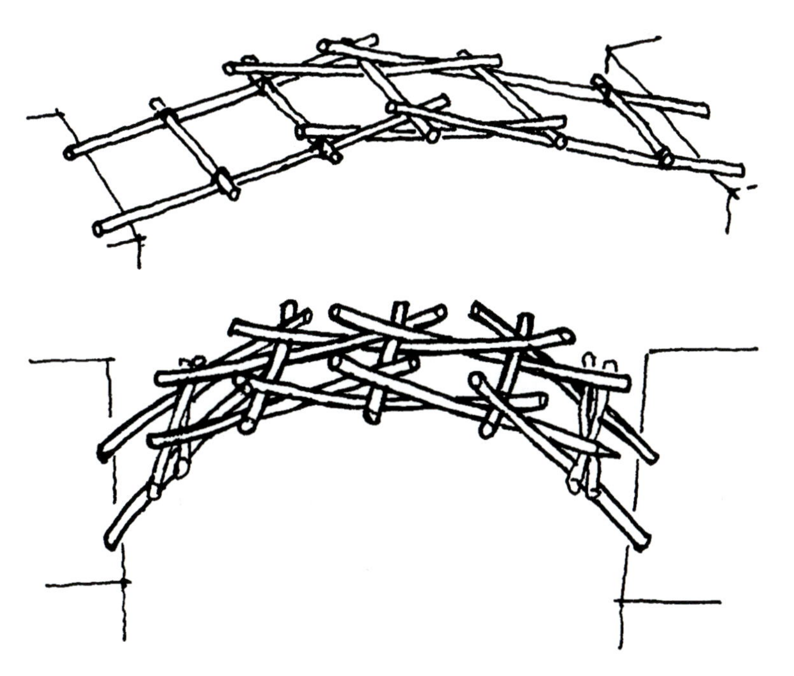

▲ **2.11** (a) and (b) Leonardo da Vinci's proposals for temporary bridges. (Sketches by A. E. Piroozfar.)

Another planar grillage was proposed in the Renaissance period by the Bolognese painter and architect Sebastiano Serlio. In 1537, Serlio published a prospectus for a treatise on architecture in seven books, and in the fifth book he proposed a planar grillage for a '. . . ceiling which is fifteen foot long and as many foot broad with rafters which would be fourteen feet long . . .' (Murray, 1986, p. 31). He notes that 'the structure would be strong enough' (Serlio, 1611, p. 57). In the fourth book, tenth chapter, Serlio makes two sketches for door frames which are also planar grillage

▲ **2.12** Pedestrian RF bridge. (Photo: Andy Tyas.)

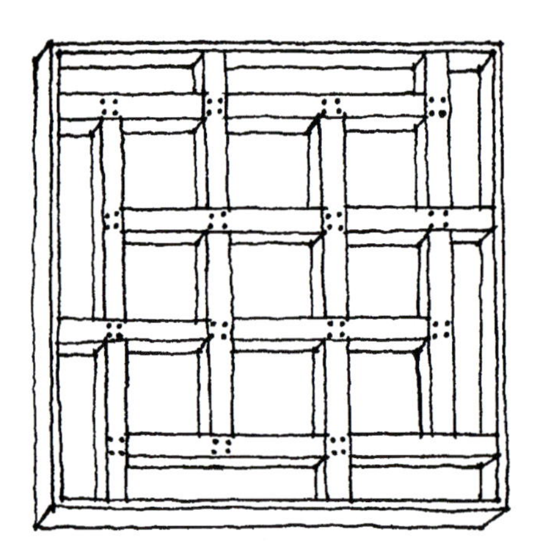

▲ **2.13** Serlio's solution for a 15-foot ceiling. (Sketch by A. E. Piroozfar.)

structures. Serlio's planar grillages are very similar to Honnecourt's solution for spanning long distances with shorter beams. Figure 2.13 shows Serlio's idea.

Less then a century later (1699), John Wallis described the inclined and planar grillage assemblies he had studied in his *Opera Matematica*. In 1652–53, while lecturing at King's College Cambridge, he built physical models of grillage structures. Wallis investigated how to span longer distances with elements shorter than the span by looking at three- and four-beam RF assemblies that had sloping beams. The multiple grillages were planar assemblies (Figures 2.14 and 2.15). It is not clear from his writings whether these structures were built on a large scale at the time, going beyond the small-scale physical models that he used for teaching and exploring the geometrical and structural principles. It is very likely that Wallis was only a scientist and researcher, fascinated by these structures which he explored in great detail, and that he was never involved in scaling them up and using them in real building structures. Despite that, his contribution is of great importance because he was the first to describe the geometry of flat grillages and to study their structural behaviour. Wallis's *Opera Matematica* is the first known written document exploring the load transfer of the structure.

Wallis also explored the different planar morphologies of grillages and worked out their geometry in order to study load paths through the structure. The assemblies are constructed by connecting elements which are notched and fitted into one another. The structures that

▲ **2.14** Three- and four-beam RF assemblies. (Sketch by A. E. Piroozfar.)

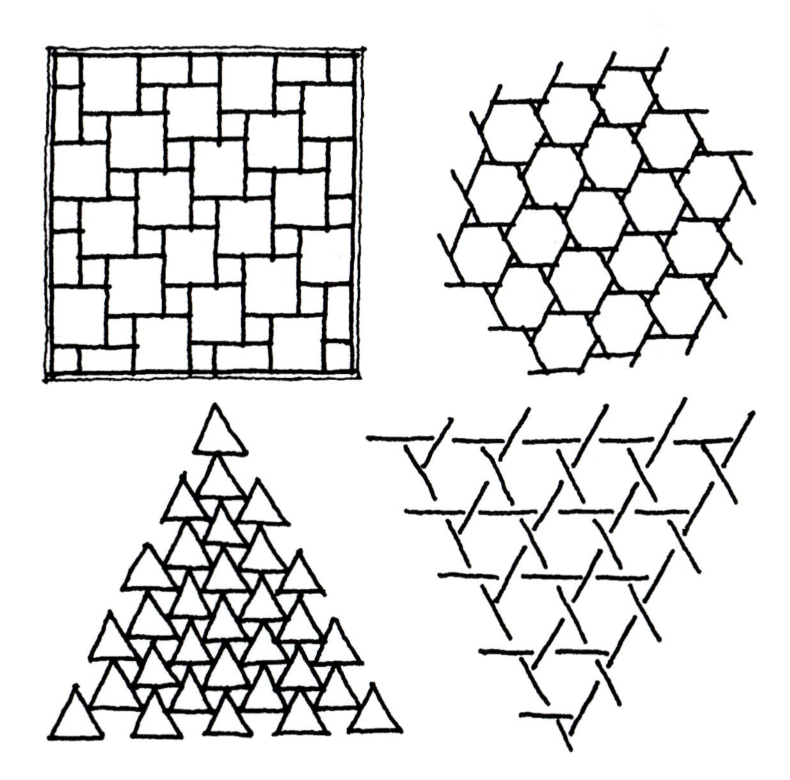

▲ **2.15** Planar morphology of grillage structures. (Sketch by A. E. Piroozfar.)

Wallis studied are very similar to Leonardo's grillage assemblies. Some examples that he studied are presented in Figure 2.15.

Other interesting historical examples of flat grillages are presented in the atlas, *Traite de L'art de la Charpenterie*, written by A. R. Emy, who was a Professor of Fortification of the Royal Military School, Saint-Cyr, and a member of the French Royal Academy of Fine Arts. It was published in Paris in 1841. Unfortunately, the book gives no information in the text about where these structures (presented in Figure 2.15a and b) were used, and the spans and sizes of the elements involved. Nevertheless, it represents further evidence of the long-term historical development of grillage structures.

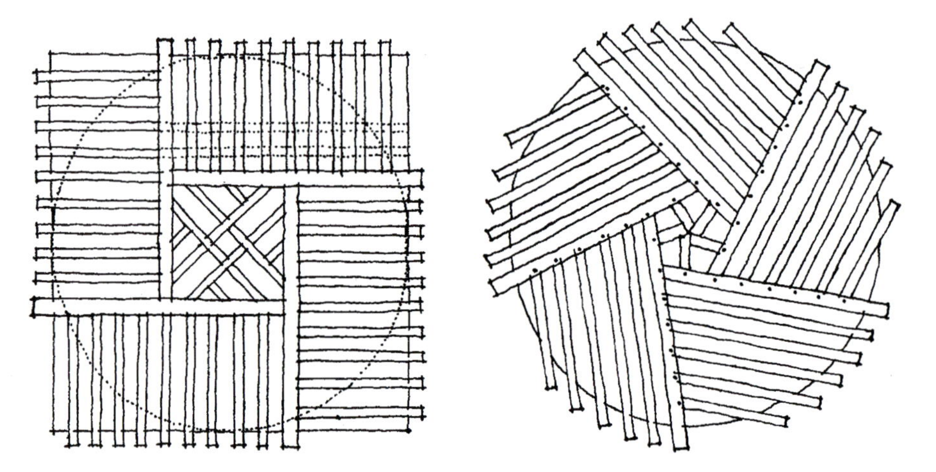

▲ **2.16** Example of a grillage structure (a) over a square plan and (b) over a circular plan. (Sketches by A. E. Piroozfar.)

▲ **2.17** Flat grillage by Serlio. (Sketch by A. E. Piroozfar.)

Thomas Tredgold, in his book *Elementary Principles of Carpentry*, devotes a whole section to 'Floors constructed with short timbers'. It is interesting to note that Tredgold (1890, p. 142) describes these ceilings as '. . . structures which can not be passed over without notice and yet are scarcely worthy of it . . .' and as '. . . more curious than useful . . .' because they are seldom applied. They are only useful when the timber is not long enough. He describes the 'Serlio-type ceiling' and gives another example designed by Serlio (Figure 2.17), as well as the research done by Dr Wallis. The main difference between the structures that Tredgold describes and the RF is that they are planar grillages in which the members are joined by mortises and tenons.

Several three-dimensional grillage structures that have a greater similarity to the RF were constructed in the twentieth century. These include the roofs at Casa Negre, San Juan Despi, Barcelona (1915) and Casa Bofarul, Pallararesos, Tarragona (1913–18), both designed by the Spanish architect Jose Maria Jujol (Flores, 1982). Inspired by Gaudi's architecture of spiral forms, such as the ceiling of Casa Battlo, Jujol designed roof structures of mutually supporting and spiralling beams. In both buildings the structures used are identical to the RF.

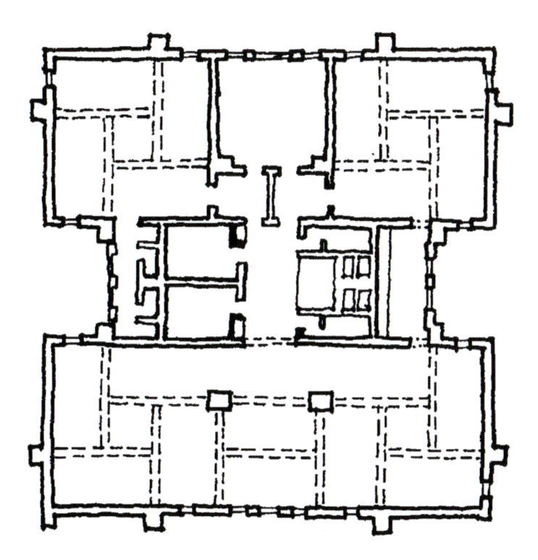

▲ **2.18** Mill Creek Public Housing Project in Philadelphia 1952–53 – plan view. (Sketch by A. E. Piroozfar.)

The floor structure used in the Mill Creek Public Housing Project in Philadelphia, designed in 1952–53 (Figure 2.18) by the architect Louis Kahn, used a four-beam planar grillage in the high-rise buildings (Scully, 1962). The main advantage of using the planar grillage in this housing project is the avoidance of columns within the plan, which consequently made it easier to plan the spatial organization of the spaces. The span is 15 metres. The configuration is identical to a planar medieval four-beam grillage. Unfortunately, this project was never realized.

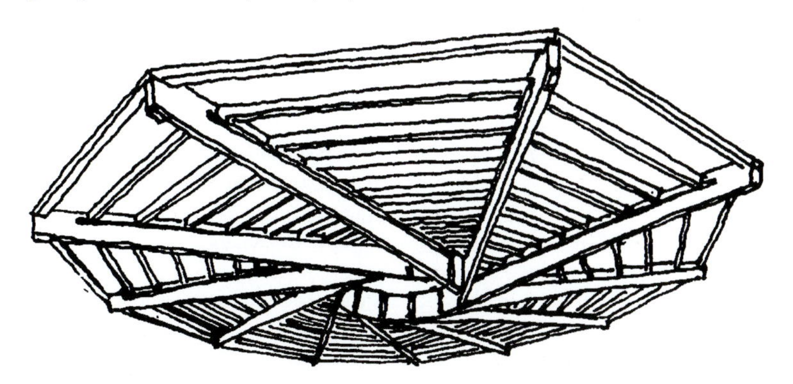

▲ **2.19** Salt storage building in Lausanne in Switzerland. (Sketch by A. E. Piroozfar.)

A more recent planar grillage structure is the roof of a salt storage building at Lausanne in Switzerland (Figure 2.19). Eleven tapered, glulam beams are used to cover the regular polygonal plan of this building, which has a span of 26 metres (Natterer, Herzog and Volz, 1991).

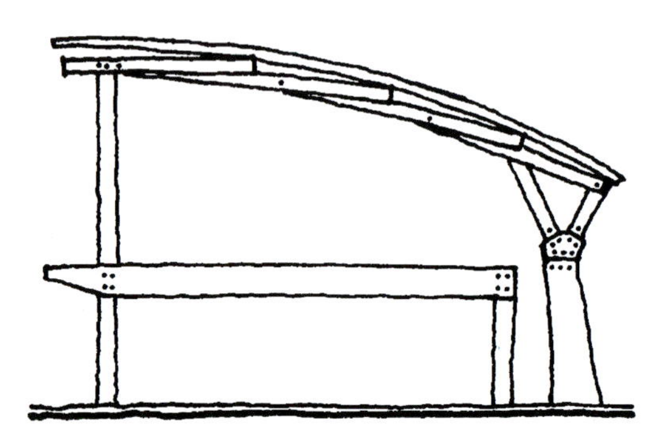

▲ **2.20** Langstone Sailing Centre - section through the interlocking timber structure. (Sketch by A. E. Piroozfar.)

Another design using a similar structure to the RF is the roof of the Langstone Sailing Centre, constructed in April 1995 (Figure 2.20). Influenced to a great extent by traditional shipbuilding technology, the Hampshire County Architect's concept was to produce a 'locked chain' effect for the roof. By use of a series of physical models, Buro Happold, who were the engineers for the project, studied the structural and geometrical implications. The roof structure is constructed of pairs of interlocking pitched pine timber members which span 10.5 metres. The members are connected by shear plate connectors hidden neatly by oversized washers. An extremely high degree of accuracy was necessary because single bolts passed through up to eight shear plate connectors and the clearance in the holes was only 2 mm (*The Structural Engineer*, 1995).

Both the Langstone Sailing Centre roof structure and Leonardo's temporary bridges are assemblies of simply supported interlocking beams, which means in practice that both types of structure 'work' in the same way. It is interesting to note that the structure has been referred to as 'unique' (*The Structural Engineer*, 1995, p. 3), although the structural principle is identical to Leonardo's structures.

More recent RF buildings that have been innovative in their use of the RF principle and integrated it architecturally in the design will be described and analysed in detail through the work of Japanese and UK designers presented later in this book. The projects present a detailed account of the design process for each scheme, as well as describing their designers' vision. Often, through the interviews with the designers (architects and engineers) and clients, the missing links which help us to

▲ **2.21** Hans Scharoun's Berlin Philharmonic reinforced concrete RF. (Photo: Peter Blundell-Jones.)

understand how and why the RF was integral to the particular design project have been established. The reciprocal frame projects include the work of Japanese designers: architect Kazuhiro Ishii with his designs for the Spinning house, Seiwa Burnaku Puppet Theatre and the Sukiya Yu house; architect Yasufumi Kijima with his design of the Stonemason Museum; and engineer Yoichi Kan with Torikabuto, the Life Sciences Laboratory. In addition, the work of UK designer, Graham Brown, who was the first to name the reciprocal frame, is presented through his designs for the Findhorn Foundation whisky barrel house and Colney Wood burial park buildings. Also, at the end of the book several recently constructed RF buildings are presented.

This account has presented only some of those structures that have been built in the past and which have some similarities to the RF. They are by no means the only examples. RFs and structures similar to them have been built by many cultures throughout history. If one tried to include all these structures the list would be beyond one book. Still, one ought to mention Hans Scharoun's Berlin Philharmonic reinforced concrete RF (Figure 2.21), the multiple grids by Gat (Figure 2.22), the Rice University bamboo canopy by architect Shegiru Ban and engineer Cecil Balmond (Figure 2.23), as well as the work of many researchers such as John Chilton, Vito Bertin, Messaoud Saidani, Olivier Baverel, Joe Rizotto and many others. The research work will be presented in more detail in the geometry and morphology chapters of this book.

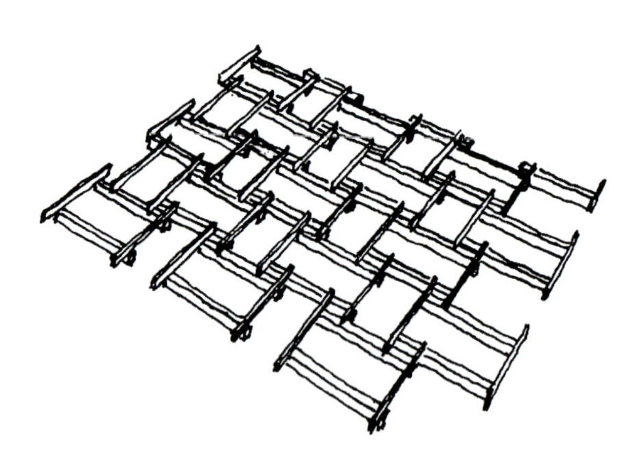

▲ **2.22** Multiple grids by Gat. (Sketch by A. E. Piroozfar.)

▲ **2.23** Rice University bamboo canopy by architect Shegiru Ban and engineer Cecil Balmond - detail. (Sketch by A. E. Piroozfar.)

This section shows that the inspiration to use RFs and similar structures in buildings has come from many different sources. Although scattered all over the world, they all contribute in their own way to the unique language of RF architecture, forming stepping stones in its history.

3 MORPHOLOGY

In the context of this book, the term morphology will be used to describe the arrangement of structural members that form the reciprocal frame to create a particular three-dimensional configuration. By varying the geometrical parameters of the reciprocal frame (RF) structure, such as the length and number of beams, radii of inner and outer polygons, and beam slopes, as presented further in the geometry section (see Chapter 4), a designer can achieve a great number of different morphologies. In addition, one has the option of using single RF units or multiple RFs, which adds to the versatility of the system and helps create different architectural expressions.

Having defined the RF as a structure made up of mutually supporting beams placed in a closed circuit, one would expect the most obvious plan form of a RF building to be circular. Indeed, most of the RF buildings constructed to date have regular polygonal or circular plans. In the case of regular plan forms, all RF members are identical, which offers the possibility of modular construction. RF designer Graham Brown uses the modular approach in most of his designs (as described in Chapter 10). This allows for higher quality and greater speed of construction.

The circular plan form is one of the first to have been used. Many vernacular buildings throughout human history (mud huts, cave dwellings and the like) had approximately circular plan forms. They would appear to have a protective, womb-like quality. Also, circular and regular polygonal forms are often used for some types of public buildings such as churches, concert halls, sports stadia and museums. However, circular and polygonal plan forms are quite rigid geometrical shapes. As such, they can be subdivided in a limited number of ways that 'work' geometrically. The spiralling effect of the RF structure, with beams offset from the centre, adds an additional constraint which predicates an obvious way of subdividing the spaces, using partitions that follow the beam lines in plan. Thus, the best applications of RFs are for open-plan functions and spaces without any internal partitions. This is not to say that partitioned spaces are impossible using the RF, only that they require more thought and care when designing them. Otherwise, the spaces may end up having odd polygonal shapes, and may be difficult to furnish and work in. When using the RF for open-plan functions, the structural expression and the RF effect can be enjoyed in totality. When looking at the RFs constructed to date (see Chapters 7–11), it becomes clear that the visual impact of the structure of self-supporting spiralling beams is very powerful.

Although circular and polygonal plan forms are the most obvious, the RF unexpectedly offers a great variety of possible RF morphologies. The RF can be used to roof any plan form: it can be used over circular, polygonal and oval but also over completely irregular or organic plan forms. Although there are no built examples of irregular RFs to date, they are a clear possibility.

▲ **3.1** Regular and irregular plan forms.

The term RF in the context of this book will be used for a structure with sloping beams that are placed in a closed circuit and are mutually supporting. However, if the definition is extended to encompass more of these units of mutually supporting members joined together, we get multiple RFs. The multiple RFs can be divided into two basic groups: multiple RF grids and complex RFs. The multiple RF grids are reminiscent of grid shells, and are formed by expanding and adding single RF units to the perimeter of the single unit to form a grid structure. Professor Vito Bertin, based at the Chinese University in Hong Kong, who researches into lever beam structures, describes these grids as generated through perimeter expansion (Bertin, 2001). Examples of multiple RF grids are Leonardo's sketches of multiple grids (Chapter 2),

Ishii's auditorium structure for the Burnaku Puppet Theatre (Chapter 7) and Kijima's Stonemason Museum (Chapter 9), as well as the recently completed laminated bamboo canopy at Rice University in the USA, designed by architect Shegiru Ban with structural engineer Cecil Balmond (Chapter 2).

The other group of RF structures that consist of more than one RF unit are complex RFs. These are formed by combining single RF units that are inserted in the central opening (the inner polygon) instead of being added around the perimeter as in the case of multiple grids. Bertin (2001) describes them as being generated through interior densification. An example of complex RFs is Ishii's Spinning House RF roof as well as his exhibition hall at the Seiwa Burnaku Puppet Theatre. The latter example, as explained in detail in Chapter 7 of this book, has a double RF unit at the outer circle consisting of RF beams spiralling clockwise and anticlockwise, creating both a beautiful and earthquake-resistant building. The double RF structure increases the structural redundancy of the roof and helps overcome the danger of progressive collapse, as explained in more detail in the section on structural behaviour (see Chapter 5).

Some explorations of RF morphology and possible architectural applications of the system with both single and multiple RF grids, as well as with complex RFs, have been carried out at the School of Architecture, University of Sheffield. The aim of the explorations was to look at the potential of the structure for creating different morphologies. By varying the parameters of the structure, a great number of original forms were created. This enormous potential for creating different RF morphologies gives the designer a unique opportunity for creating a new expression with each different RF configuration.

The research at the University of Sheffield was designed to explore the potential for creating different forms of RFs and how they may be used in architecture. In order not to constrain the opportunities of morphology, structural behaviour and connection detailing were not considered at this stage. The presented images of single RFs, multiple RF grids and complex RF configurations explored through physical modelling and sketches pre-sent some of the possible forms that can be created. If these were to be used in building design they would need to be developed further. The forms would need to be rationalized to achieve efficient structural design. In addition, depending on the material chosen for the structure, appropriate joining details would need to be designed. These issues are discussed further later in the book, in the section on structural behaviour (see Chapter 5).

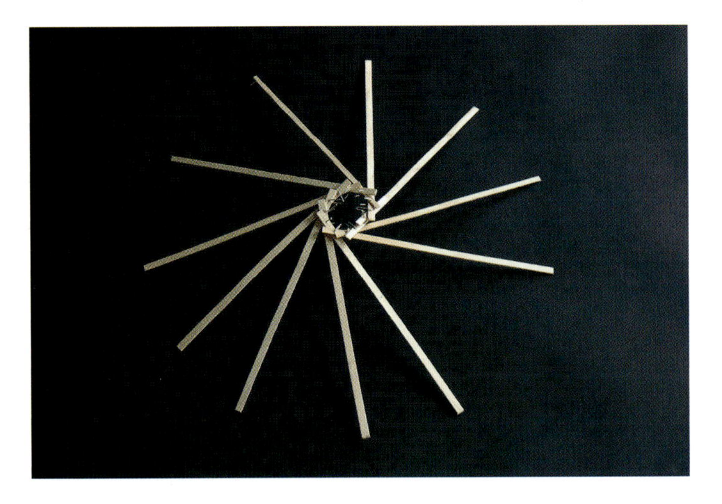

▲ **3.2** Single RF structure with 11 beams.

▲ **3.3** Single RF structure with four beams.

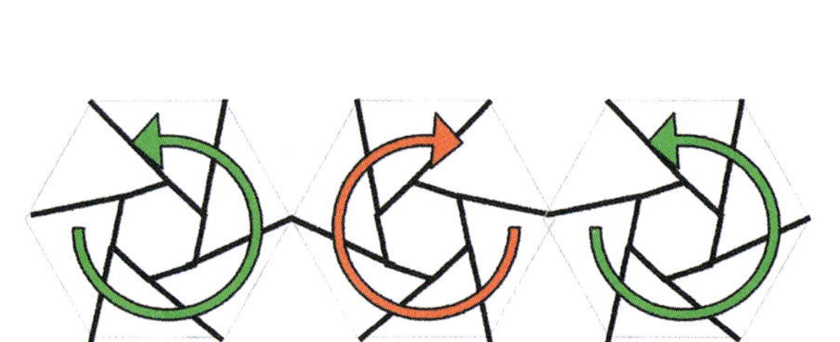

▲ **3.4** Single RF units with clockwise and anticlockwise beams.

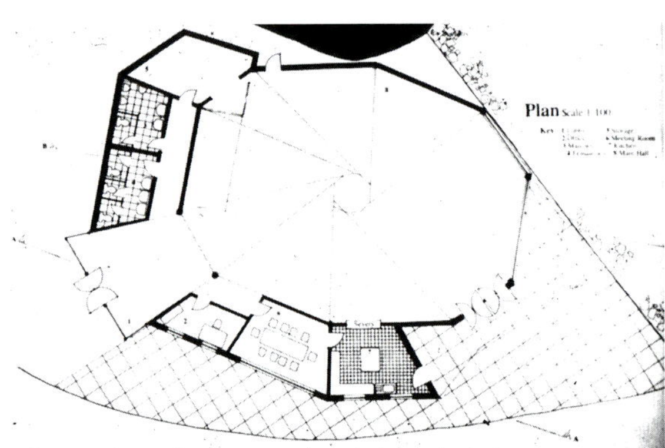

▲ **3.5** Community hall design – plan view.

▲ **3.6** Community hall design.

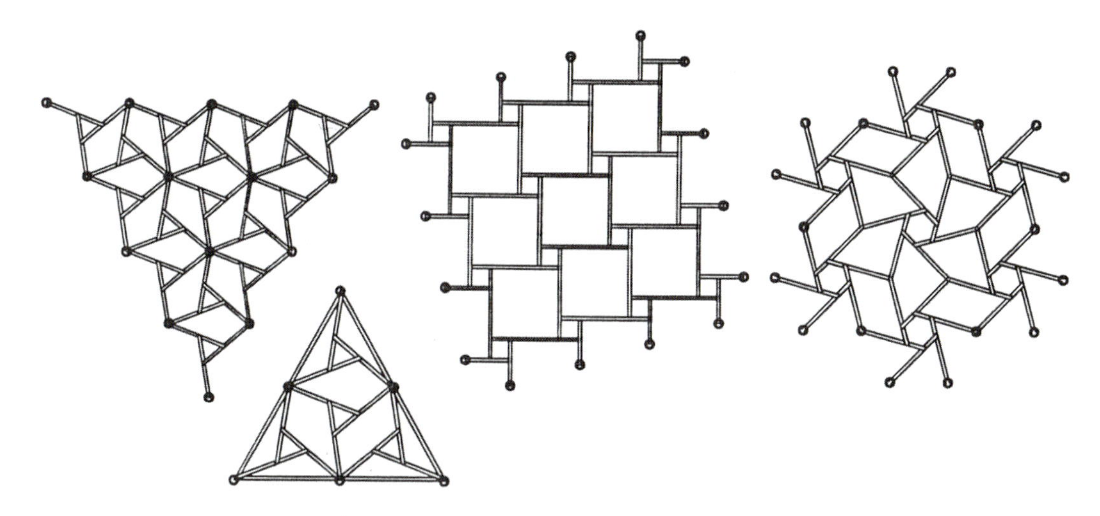

▲ 3.7 Examples of RF multiple grids.

▲ 3.8 Multiple RF grid consisting of single four-beam RFs.

▲ **3.9** Multiple RF grid consisting of single four-beam RFs – detail.

▲ **3.10** Sheffield architecture students constructing a multiple RF grid dome – 1.

▲ **3.11** Constructing a multiple RF grid dome – 2.

▲ **3.12** Constructing a multiple RF grid dome – 3.

▲ **3.13** Constructing a multiple RF grid dome – 4.

▲ **3.14** Constructing a multiple RF grid dome – 5.

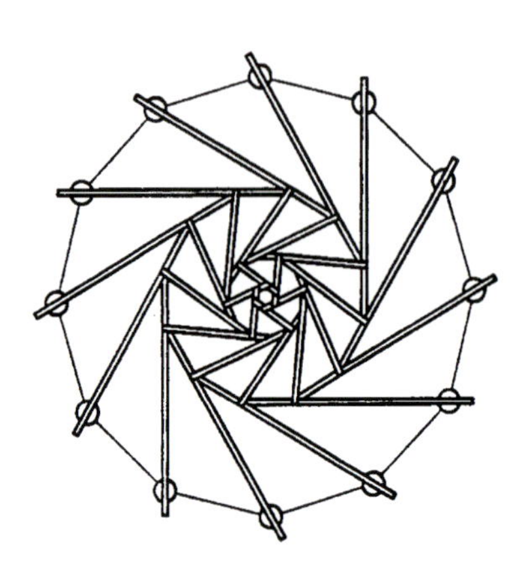

▲ **3.15** Example of a complex RF.

▲ **3.16** Constructing a complex RF – 1.

▲ **3.17** Constructing a complex RF – 2.

▲ **3.18** Student explorations – bridge design.

▲ **3.19** Student explorations – bridge in context.

▲ **3.20** Student explorations – bridge detail.

▲ **3.21** Student explorations – da Vinci-like bridge design.

▲ **3.22** Student explorations – da Vinci-like bridge design detail.

▲ **3.23** Student explorations with RF grids – 1.

▲ **3.24** Student explorations with RF grids – 2.

▲ **3.25** Student explorations with complex RFs.

▲ **3.26** Student explorations with complex 3D RFs.

▲ **3.27** Student explorations with complex 3D RFs – detail.

▲ **3.28** Sheffield student explorations with visiting Professor Tony Hunt and the author.

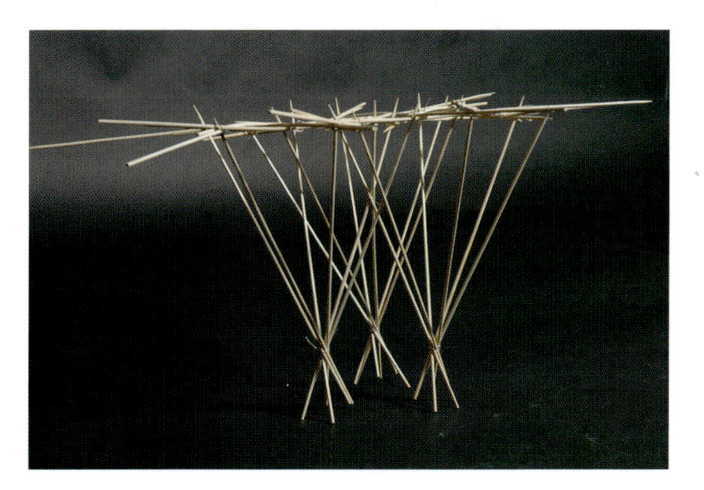

▲ **3.29** Student explorations with complex 3D RFs.

▲ **3.30** Student explorations with complex 3D RFs – detail.

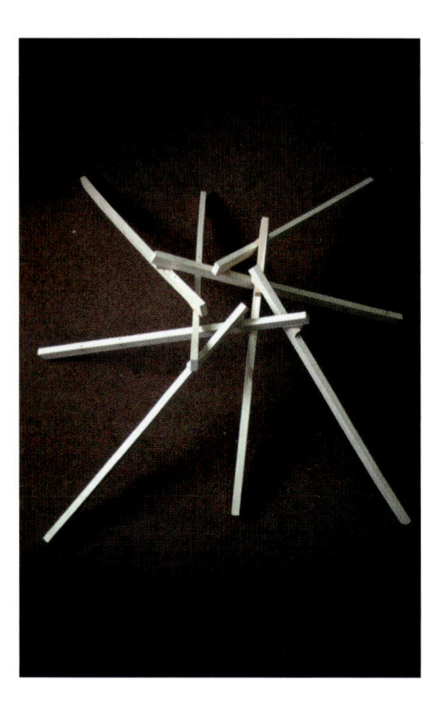

▲ **3.31** Simple RF with rotated beams.

There have been a number of researchers who, over the last 10–15 years, have investigated different aspects of RF structures. In the field of RF morphology, the work of Dr John Chilton, Professor at the Lincoln School of Architecture, who, with several of his research students pioneered research into RF morphology, geometry and structural behaviour, should be mentioned. Together with masters student Orlando Ariza, he investigated the relationship of multiple RFs and polyhedra. In addition, through both small-scale physical models and computer simulations, they investigated possible applications of complex RF grids. Being both an academic as well as a practising structural engineer, Chilton has been the consultant engineer for most of Graham Brown's RF designs (as described in Chapter 10). In an interview carried out for the purpose of this book, he stated:

> *'With the RF it is all about the roof structure. There should be more explorations with adventurous forms, woven structures, basket woven forms. The RF is like a collapsed tensegrity structure. The tension and compression members are the RF beams – they work in bending, thus they replace the tension and compression members. More explorations need to be done.'*

He adds:

> *'The complexity of the geometry in my view is the main reason why the RF has not been used a lot. However, with modern computers this stops being a problem. It is a special structure. Circular and square buildings with RFs work well. Other forms have not been explored enough.'*

In parallel and more or less simultaneously with the research of John Chilton is the exploratory work into RF morphology by Professor Vito Bertin of the Chinese University of Hong Kong. Bertin refers to and classifies RFs as part of a broader group of so-called 'mutually supported stick structures'. The subgroup that he calls 'lever beam structures' are identical to RFs. The name 'lever beam structures' comes from the way these structures work and transfer the load. His investigations look at the parametric relationships of the RFs with an aim of producing a catalogue of possible form variants. Part of his research is carried out through the construction and testing of large-scale physical models. Through a physical model of a shallow dome 10 metres in diameter, constructed from bamboo rods 1.5 metres long and 4 cm in diameter, he establishes several interesting facts. During construction of the dome, consisting of triangular and hexagonal RFs, the beams were tied with plastic ties. He found that when the RF dome was complete the ties were not needed for most of the joints in the upper portion of the dome, where friction forces kept the bamboo sticks from sliding. In addition, after the load testing using distributed weights of 15 kg attached to 20 locations, he noticed that the outward thrust of the perimeter anchoring members was so small that the friction of the sticks on the grass-covered ground was enough to prevent movement. When the load was increased to test for its failure point, it was noticed that when some members failed through buckling because of excessive bending, a hole was created in the dome but the dome did not collapse. This showed the inherent capacity for load sharing of the structure and its ability to redistribute forces. An image of this structure is presented in Figure 3.32.

Another researcher who has carried out research on multiple RF grids is Olivier Baverel, who is a lecturer at the School of Architecture in Nancy in France. His Ph.D. research work, supervised by Professor Nooshin of Surrey University, investigated the complex geometrical forms that can be created by using multiple grids. By carefully combining the number of members in single units, their inner radius and the length of members, one can control the curvature of the complex structure. Baverel, through his research, defined the combinations necessary to obtain the form of a dome, cone or doubly curved grid. He uses genetic algorithms to generate and define the complex geometrical relationships between members and units in the grid. Baverel refers to these structures as 'nexorades', the name coming from the Latin word '*nexor*', which means link. His work includes the construction of large-scale models of nexorades. He uses flexible tied joints for his models, which allow for the members to adjust and rotate until they find a stable configuration. If used for real buildings the joint design would need to be altered to offer more secure connections. Further work would need to be carried out on the design of the connections before the multiple grids, or 'nexorades', could offer a viable building solution. Baverel's work is a very valuable contribution in the field of geometry and morphology generation of multiple RF grids.

In parallel to Baverel's work on multiple RF grids, but with different emphasis, several other researchers have been investigating the possibility of using these structures in building design. It is worth mentioning the work of Messoaud Saidani and Joe Rizzuto of Coventry University, who have done research work into the structural behaviour of similar structures. They refer to them as 'mutually supported elements' or 'MSE'. The special value of this work is that, through his Ph.D. research, engineer Joe Rizzuto, under the supervision of Messaoud Saidani, not only investigated the geometry, but in addition looked at the structural behaviour of multiple RFs, comparing the results of the structural analysis with tests carried out on physical models. For the purpose of his research, he constructed a dodecahedric dome consisting of three- and six-membered RFs. It would be interesting to continue this research and to investigate the structural behaviour of other multiple configurations.

▲ **3.32** Bertin's 10-metre bamboo multiple RF grid dome. (Photo: Vito Bertin.)

▲ **3.33** Baverel with his physical model of a multiple RF grid sphere (Photo: Olivier Baverel.)

▲ **3.34** Baverel's multiple RF grid sphere – detail. (Photo: Olivier Baverel.)

Multiple RF grids have not been broadly adopted and very few buildings that use them have been constructed to date. With the exception of the pressed laminated bamboo canopy structure at Rice University in the USA, designed by architect Shegury Ban with structural engineer Cecil Balmond, there are hardly any other examples used in building design. Yet multiple RF grids offer the possibility of creating amazing and unexpected three-dimensional shapes. All the researchers mentioned continue their research into multiple grids. Hopefully, through their work and the work of others, multiple RFs will become a more viable practical option in building design in the future.

4 GEOMETRY

A reciprocal frame is a three-dimensional structure with complex geometry. Understanding the geometry of the structure and the parameters that define it is important in order to make it possible to design and construct a reciprocal frame (RF) building. The parameters that define RF units with regular polygonal and circular geometry are the following:

- Number of beams (n)
- Radius through the outer supports (r_o)
- Radius through beam intersection points (r_i)
- Vertical rise from the outer supports to the beam intersection points (H)
- Vertical spacing of the centrelines of the beams at their intersection points (h_2)
- Length of the beams on the slope (L).

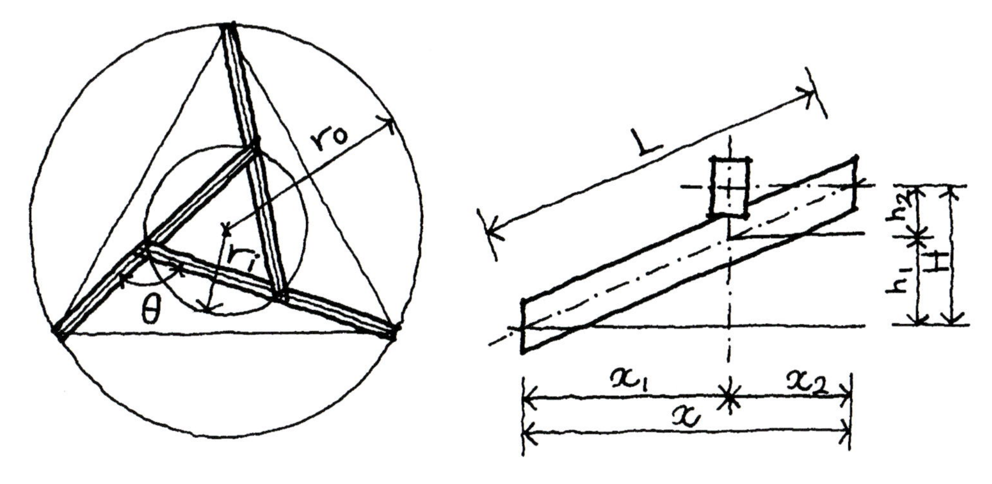

▲ **4.1** Geometrical parameters for RF structures. (Sketch by A. E. Piroozfar.)

The parameters that define the geometry of the RF, and their interdependence derived using basic trigonometry as proposed by Chilton and Choo (1992), can be determined from equations (4.1)–(4.7). In these equations, θ is the sector angle between the beams (that is, the angle between the beams when viewed in plan), x is the overall length of a beam in plan, and x_1 and x_2 are the plan length to the first intersection and plan length between intersections.

$$\theta = \frac{360}{n} \tag{4.1}$$

$$x_2 = 2r_i \sin\frac{\theta}{2} \tag{4.2}$$

$$x_1 = \left[r_0^2 - \left(r_i \cos\frac{\theta}{2}\right)^2\right]^{\frac{1}{2}} - \frac{x_2}{2} \tag{4.3}$$

$$x = x_1 + x_2 \tag{4.4}$$

$$h_1 = H\frac{x_1}{x} \tag{4.5}$$

$$h_2 = H - h_1 \tag{4.6}$$

$$L = (x^2 + H^2)^{\frac{1}{2}} \tag{4.7}$$

Variation of the Parameters

Research (Chilton, Choo and Popovic, 1995; Popovic, 1996) has been carried out investigating the impact of the variation of the main RF parameters. The effect of varying the spacing of the beam centrelines at their intersections on the depth of the beam or truss cross-section has been examined. Special emphasis has been given to the impact these variations may have on the physical construction of the RF.

For instance, where h_2 is equal to or less than the depth of the solid beams used in an RF, the upper beam is usually notched on its underside so that the desired vertical beam spacing can be obtained. The size of the notch also depends on the width of the beams and their angle of inclination. The notch weakens the upper beam at a point of high shear and can necessitate reinforcement of the joint, as in the case of the 13-metre-diameter modular RF house at Saorsa, Ardlach, Nairn, Scotland, designed and constructed by Graham Brown.

On the other hand, where h_2 is small (or zero) it is easier to connect the supported beam onto the side of the supporting beam at the intersection points. In this way, and when the beams are horizontal, a planar

RF structure is formed, similar to the medieval examples discussed in detail in Chapter 2.

In cases where h_2 is large, the beam or truss depth may have to be increased solely so that the RF members come into contact at the point where they cross. Alternatively, packing pieces or stud columns would be required to transfer loads between the primary structural elements at the intersections.

In practice, all this means that a set of well-chosen ratios of RF parameters needs to be decided upon to form the three-dimensional RF structure. If, for example, five beams are used for an RF with a ratio between inner and outer radii of 0.3 (a structure with, for example, an outer radius of 8 m and a central opening radius of 2.4 m) and a rise of 2 m from the outer supports to the inner polygon, the required vertical spacing between the beam centrelines is 0.615 m.

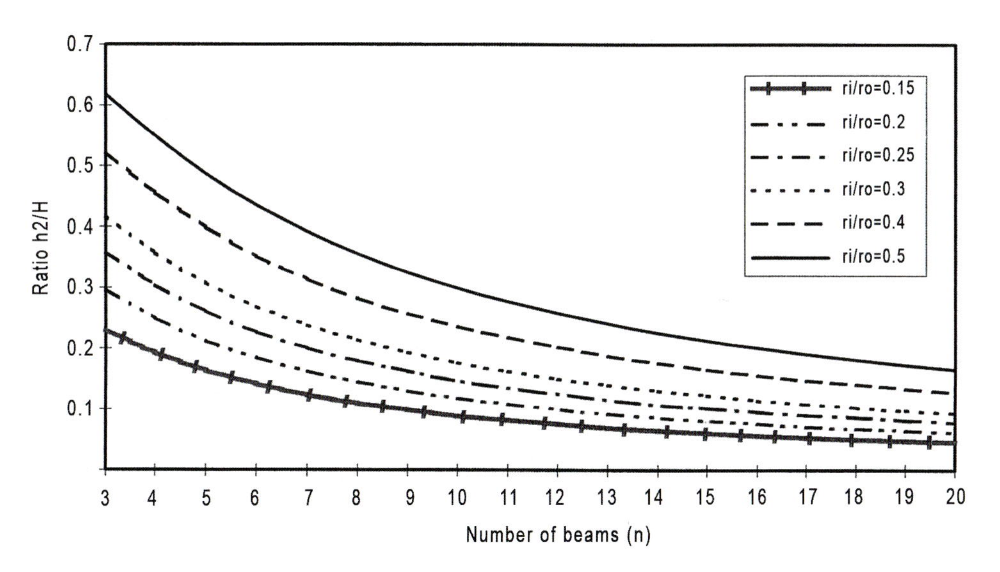

▲ **4.2** Relationship between the ratio h_2/H and the number of beams for different ratios of r_i/r_o.

The graph in Figure 4.2 is a convenient tool for preliminary design of RF structures. Given the plan dimensions (r_i and r_o) and the rise H of a regular polygonal RF roof, the curves that have been derived using equations (4.1)–(4.7) can be used to select the most appropriate number and/or depth of beams for the structure. In the graph, the h_2/H ratio gives the distance between the beam centrelines. If this distance is too great, the beams will not touch. Alternatively, for a distance that is too small, the notches at the support points of the RFs become too deep. The graph shows that not all combinations of parameters form RFs that work geometrically, and to make an RF work we have to use an appropriate combination of parameters. In this discussion, only geometric parameters have been considered. However, the ratio of r_i/r_o affects the perception of the building architecturally and also, from the structural point of view, the distribution of shear forces and bending moments in the beams of the RF. The section on structural behaviour (Chapter 5) discusses these issues in more detail. It is important to note, however, that the RF is a complex structure and all factors should be considered simultaneously. The consideration of only the geometrical parameters in isolation would be very misleading.

Further investigation of the RF geometry through parametric studies has been carried out using specialist parametric study software, Digital Project developed by Whitbybird. The summary of the findings is presented on the spreadsheet in Figure 4.3. It shows a number of combinations of RF parameters derived using equations (4.1)–(4.7) described earlier. For the purpose of this investigation only regular RFs have been considered. The size of the RF beams is normally determined by the loads on the roof (self-weight, wind, snow and in some circumstances earthquake loads), as well as the distance they need to span, as will be explained further in the section about structural behaviour (Chapter 5). However, in order to carry out a study of the geometry only, in the spreadsheet the RF beam depths chosen are a proportion of the span (span/15).

The parametric studies showed that some combinations of parameters are better than others. In the spreadsheet (Figure 4.3) the roof slope has been varied to create a maximum number of possible combinations. The combination of parameters shaded in red in the table are geometrically impractical. In this particular case it is because there are too many RF beams coming together at the inner opening, which cannot be accommodated physically because of the fixed size of the inner opening. For the same reason, RFs with very steep beams (with an angle exceeding 50 or 60 degrees) would be difficult to construct. Figures 4.4–4.10 show some of the 'appropriate' combinations of RFs, whereas Figures 4.11–4.14 show some 'impractical' combinations.

The parametric studies also indicated that the steeper the roof pitch, the smaller the inner radius needs to be, while maintaining contact between mutually supporting beams (Figures 4.15–4.19). It follows that for a roof with a lower pitch, the inner opening would have to be larger.

Number of Rafters		Beam Depth (D)	200 mm × 60 mm			400 mm × 90 mm			600 mm × 120 mm			800 mm × 180 mm		
		S(o)	Span 3 m r(o) = 1.5 m			Span 6 m r(o) = 3 m			Span 9 m r(o) = 4.5 m			Span 12 m r(o) = 6 m		
		S(i)	0.5 m	1 m	1.5 m	1 m	2 m	3 m	2 m	4 m	6 m	2.5 m	5 m	8 m
	3	Roof Pitch °	30.3°	16.5°	11.7°	30.3°	*1–16.5°	11.7°	24.2°	13.3°	9.8°	24.6°	13.9°	9.8°
		Rafter Pitch °	27.1°	13.1°	8.4°	27.1°	*1–13.1°	8.4°	20.7°	9.8°	6.5°	21.3°	10.5°	6.5°
	4	Roof Pitch °	32.1°	*2–18.8°	13.1°	35°	18.8°	13.1°	25.9°	14.8°	10.5°	20.6°	15.5°	10.5°
		Rafter Pitch °	29.5°	*2–15.7°	10.3°	32.2°	15.7°	10.3°	23°	11.8°	7.8°	17.6°	12.5°	7.8°
	5	Roof Pitch °	38.8	21	14.9	40.8	*3A–21.4	15	30.8	*3B–16.8	11.5	32.2	17.9	11.3
		Rafter Pitch °	36.5	18.4	12.4	38.4	*3A–18.7	12.5	28.2	*3B–14.2	9.4	29.6	15.2	9.2
	6	Roof Pitch°	45.3	24.2	16.7	46.8	*4–24.6	17	36.9	19.1	13.1	39.3	20.2	12.9
		Rafter Pitch °	43.3	21.8	14.6	44.8	*4–22.2	14.9	34.5	16.9	11.5	37	17.8	11.3
	9	Roof Pitch °	81.6	33.7	22.8	76.9	35.6	22.8	53.4	25.8	16.2	59.5	*5–27.7	16.2
		Rafter Pitch °	81.3	32	21.7	76.3	33.9	21.7	52	24.5	16	58.2	*5–26.3	16
	12	Roof Pitch °	78.1	82.8	29.9	79.2	45.8	29	79.8	*6–34.4	20.6	77.4	36.7	16.7
		Rafter Pitch °	77.7	82.6	29.8	78.9	44.8	28.9	79.4	*6–34	21.9	77	36.1	16.6

Variable parameters

S(o): Outer diameter of roof — S(o) = 2 × r(o)

S(i): Inner diameter of roof (diameter of central rooflight) — S(i) = 2 × r(i)

(shaded) Cannot achieve notch – end of beams clash

D: Depth of rafters: D = S(o)/15

Number of rafters, arranged symmetrically and equally spaced around central opening/rooflight.

Pitch of rafters determined by set parameters.

▲ **4.3** Spreadsheet showing the parametric study combinations. (Compiled by Chris Dunn.)

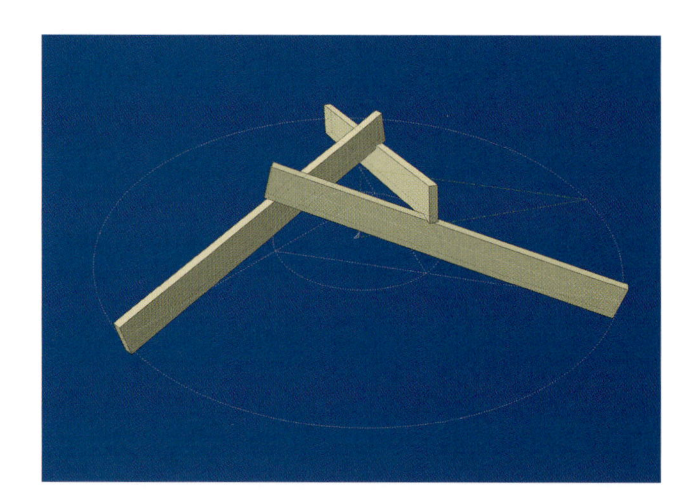

▲ **4.4** RF with three beams. (Drawing: Chris Dunn.)

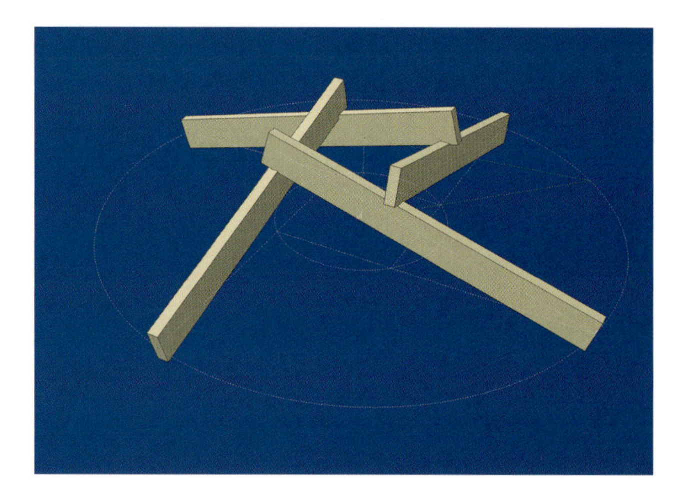

▲ **4.5** RF with four beams and small inner radius. (Drawing: Chris Dunn.)

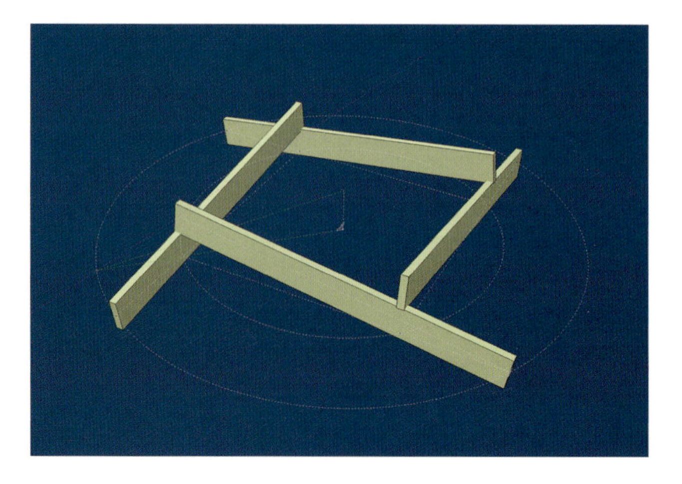

▲ **4.6** RF with four beams and large inner radius. (Drawing: Chris Dunn.)

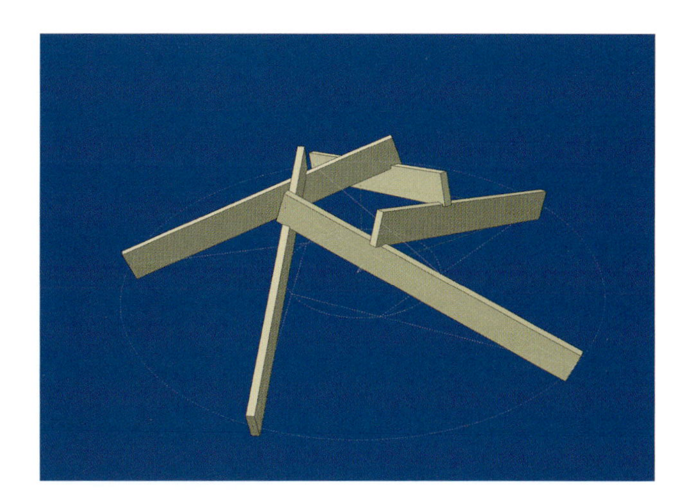

▲ **4.7** RF with five beams and small inner radius. (Drawing: Chris Dunn.)

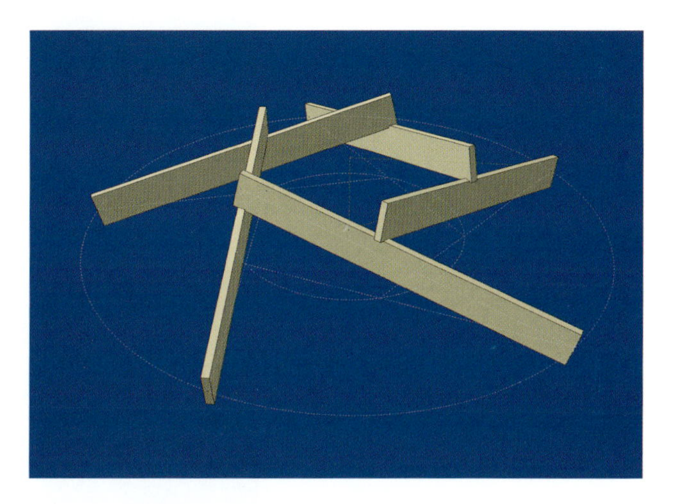

▲ **4.8** RF with five beams and large inner radius. (Drawing: Chris Dunn.)

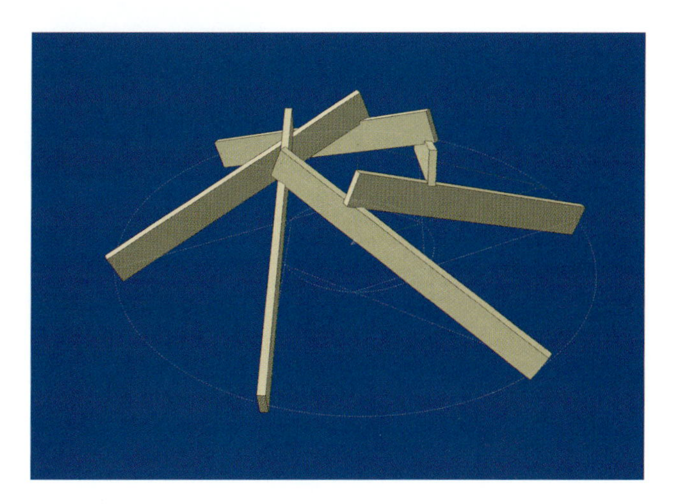

▲ **4.9** RF with six beams. (Drawing: Chris Dunn.)

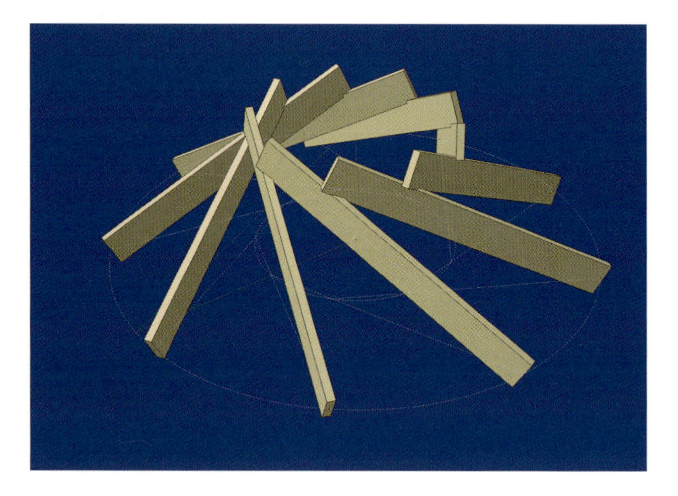

▲ **4.10** RF with nine beams. (Drawing: Chris Dunn.)

▲ **4.11** RF with 12 beams and small inner radius. (Drawing: Chris Dunn.)

This indicates that it would be very difficult to have steep roof design with a large number of beams, because the steep roof has to have a relatively small inner opening, as it would be impossible to physically fit the RF beams in the small inner opening (as seen in Figure 4.14).

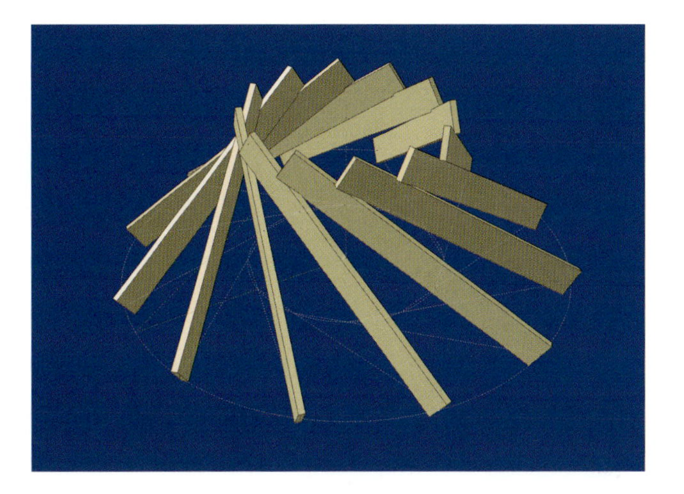

▲ **4.12** RF with 12 beams and large inner radius. (Drawing: Chris Dunn.)

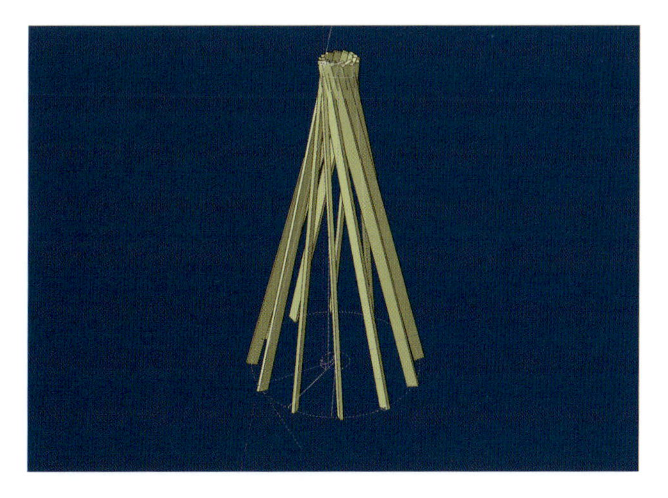

▲ **4.13** Steep RF with 12 beams. (Drawing: Chris Dunn.)

▲ 4.14 Steep RF with 12 beams – detail. (Drawing: Chris Dunn.)

In practical design, the first step towards designing an RF building would be to think about the architectural requirements for the size of the spaces, which will determine the RF spans; then to consider the cladding materials that would determine the roof slope and influence the roof weight; to choose the material for the RF structure, as whether the structure is exposed in the spaces or not will influence the type of detailing of joints (notched or otherwise if timber is used); and then the required size of the inner circle (as a roof window or not). These architectural considerations will need to be reassessed after the concept of the structure (form of the structure, number of RF beams and so on) as well as the structural analysis and detailed structural design are completed. Finally, a designed structure which fulfils both the architectural and structural requirements will only be viable if at the same time it is geometrically possible as well. It is vital when designing an RF building to

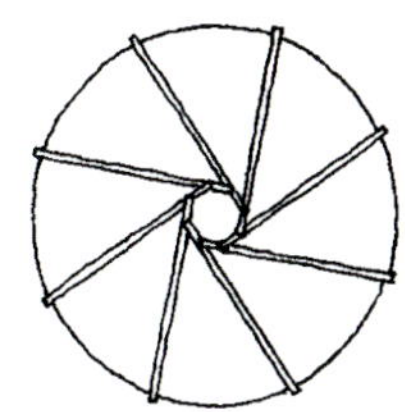
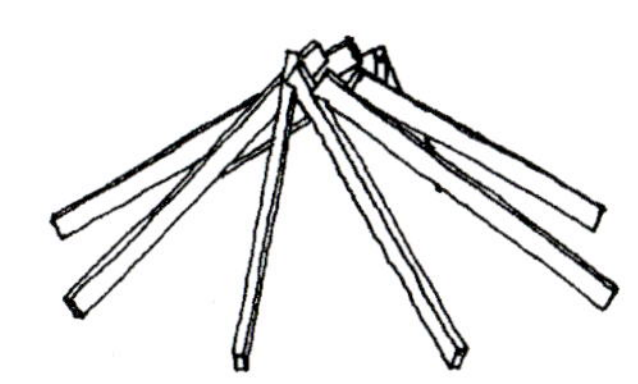

▲ **4.15** Relationship between the slope of the roof and the inner radius – 1. (Sketch by A. E. Piroozfar.)

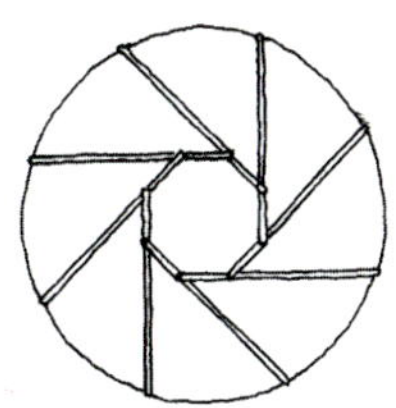
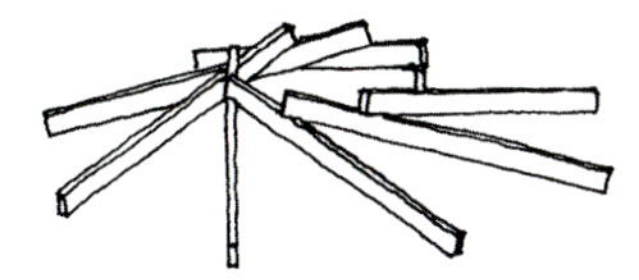

▲ **4.16** Relationship between the slope of the roof and the inner radius – 2. (Sketch by A. E. Piroozfar.)

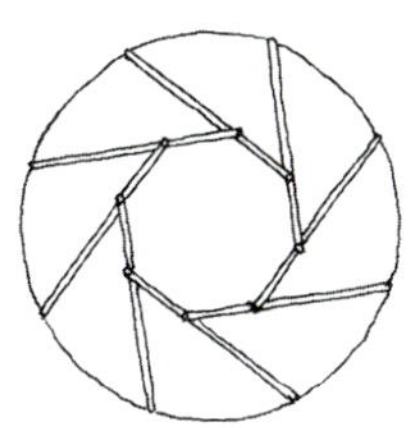
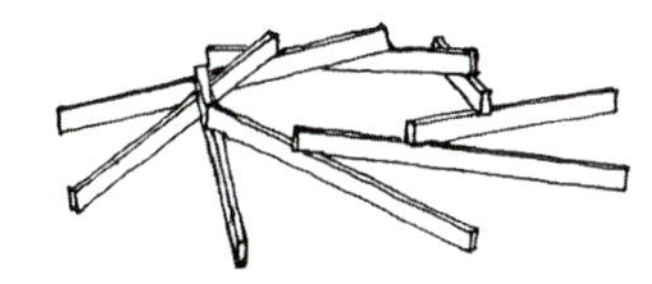

▲ **4.17** Relationship between the slope of the roof and the inner radius – 3. (Sketch by A. E. Piroozfar.)

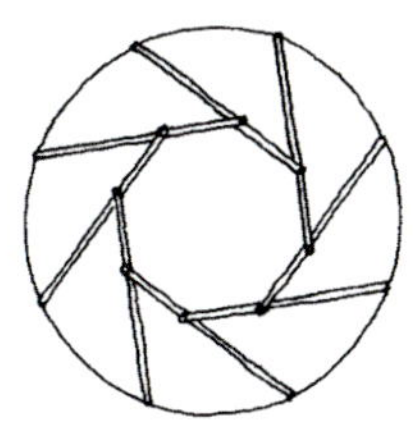
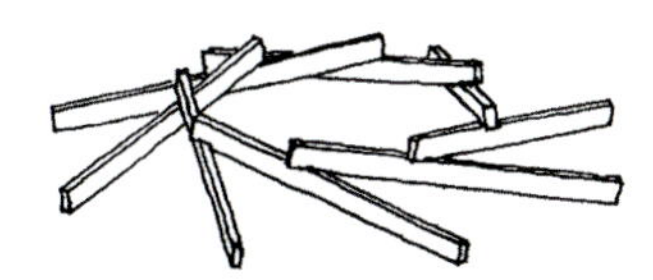

▲ **4.18** Relationship between the slope of the roof and the inner radius – 4. (Sketch by A. E. Piroozfar.)

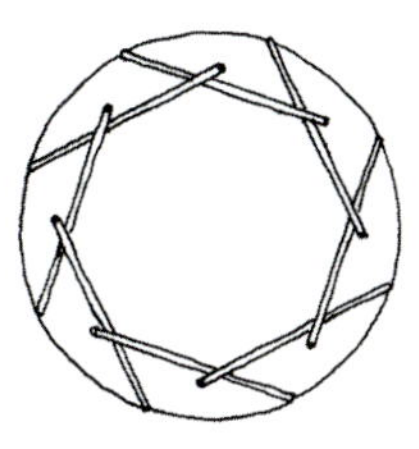
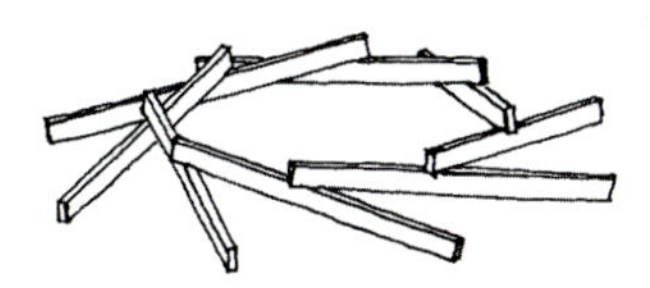

▲ **4.19** Relationship between the slope of the roof and the inner radius – 5. (Sketch by A. E. Piroozfar.)

choose realistic combinations that encompass all factors. Otherwise, there would be no roof over our RF building!

Other RF Geometries

This discussion of the geometry of the RF has so far considered only regular polygonal plan forms. If some of the conditions of regularity are relaxed, unexpected and diverse plan forms can be obtained. For example, if all the beams are not of the same length, the outer and inner polygons need not be regular, and the angles between the beams can also be different. Some very interesting irregular RF morphologies can be developed (as presented in the morphology section of this book). One can argue that the irregular forms have greater architectural potential and are potentially more interesting. However, the practical complexity of designing and constructing them would also be greater. Also, the geometry becomes more complex than when regular RF structures are used. In order to define the geometry of these structures, the irregularity needs to be described for each individual case.

All of these configurations, both regular and irregular, symmetrical and asymmetrical, show the great variety of plan geometries which can be

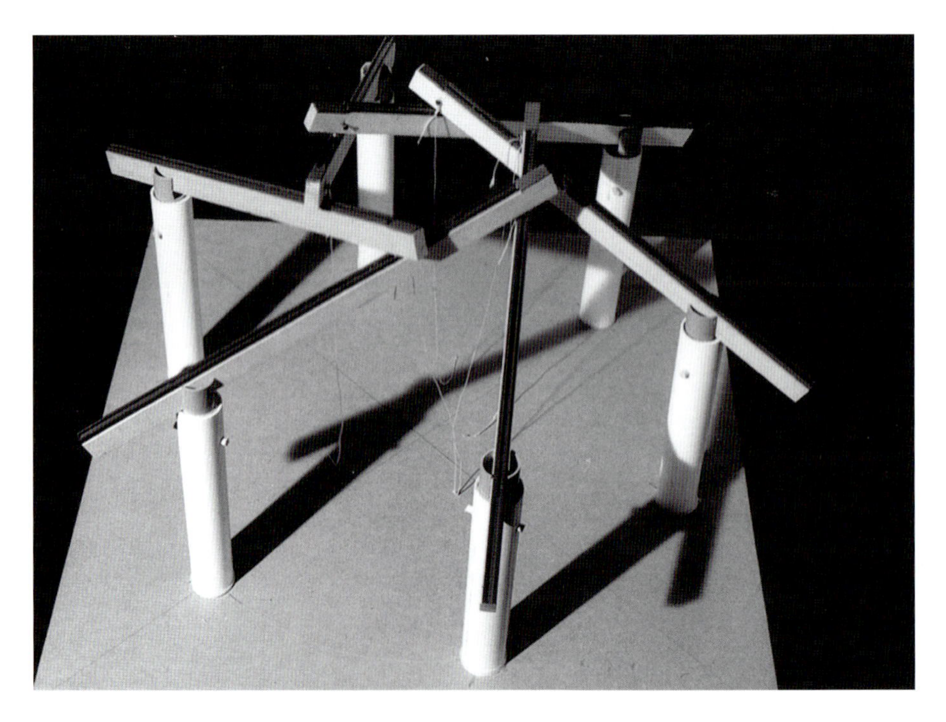

▲ **4.20** Physical model of a retractable RF structure.

obtained with the RF. This undoubtedly adds considerably to the architectural potential of this structure. In addition, as presented earlier in more detail in the morphology section of this book, it is possible to create multiple and complex RF configurations (see Chapter 3).

It is helpful that CAD tools have now been developed to such a degree that it is possible to consider very complex structures. In regular RFs both in simple, but more so in multiple and complex, units there is a great deal of repetition, which to some degree simplifies the complexity of the design task – for example, by limiting the number of different details.

Another possibility with the RF structure is the potential of creating retractable RF roofs, because the beams of the RF in plan remind one very much of the lines forming the iris of a camera shutter: some earlier investigations started by Chilton, Choo and Coulliette (1994), who have described the geometry of retractable RF structures, indicate that this is a possibility. There are no full-scale retractable RF structures constructed to date because there are many practical challenges to overcome before they become a viable building option. In order to make that happen, research needs to be extended to study how to make the cladding of the roof retract or fold with the movement of the roof. This will be a complex issue to resolve for any retractable structure. There is an additional issue that adds complexity to retractable RF structures, namely the roofing material and structure covering the inner opening of the RF roof. The inner opening roof would need to move and retract with the main structure of the RF and remain stable at every stage of the process. It is clear that it would be possible to resolve this issue technically, but it it would require some effort. This is perhaps why RFs have not been built to date as retractable structures.

List of Symbols

a, x_1	plan length of the beam from a perimeter support to lower intersection
b	plan length of the beam from perimeter support to high intersection
d	depth of beam
H	vertical rise from the outer supports to the beam intersection points
h_1	rise to first intersection
h_2	vertical spacing of the centrelines of the beams at their intersection points
L	length of the beams on the slope
n	number of beams
r_i	radius through beam intersection points

r_o	radius through the outer supports
s	distance between perimeter supports
x	overall length of a beam in plan
x_2	plan length from first to second intersection
α	horizontal angles
α_n	angle that beam *n* makes with the *x*-axis
β	beam slope angle
θ	sector angle between the beams (angle between the beams when viewed in plan)

5 STRUCTURAL BEHAVIOUR

Reciprocal frames consist of linear members which are mutually supported and interlocking, forming either a flat, horizontal structure or a pitched three-dimensional frame system. Unless stated otherwise, we normally refer to a structure with sloping beams as a reciprocal frame.

The simplest form of reciprocal frame (RF) is a beam system arranged around a single, central circle, forming a single-unit RF system. More complex forms of RFs, as explained in more detail in the chapters on morphology and geometry in this book, are multiple RFs and RF grid structures (see Chapters 3 and 4).

The minimum number of main beam members required to make the single RF structure work is three. Each member is supported at the outer end by a ring beam or a column and at the inner end is supported by the adjacent member. When the RF members are arranged regularly around a central point of symmetry, we get a regular RF structure. On the other hand, single or multiple RF structures irregularly arranged are also possible (see Chapter 3). The examples analysed structurally in this part of the book are all symmetrical structures.

RF STRUCTURES WITH INCLINED MEMBERS

An RF structure with inclined main members forming a pitched roof will typically have the inner end of the beams, or the central sections, at a higher level than the outer end sections that are at the perimeter of the structure. Arranged in this way, the members will be able to transmit the vertical forces (their own weight and any imposed loads) to the supports at the perimeter of the structure through compression in each member. For a symmetrical load (for example, self-weight) the forces in each member will be identical. However, the members will also be subjected to bending moments and shear forces, and will have to resist these forces in addition to the axial force.

The compression force must be resisted at the perimeter supports. This is often done by introducing a perimeter ring beam that can resist the horizontal thrust that will try to spread the supports and deform the structure. Examples of this type of RF frame are the Seiwa Bunraku Puppet Theatre exhibition hall by architect Kazuhiro Ishii and the RF design of the New Farmhouse at the Life Science Laboratory designed by Yoichi Kan. These are described in detail in Chapters 7 and 8.

Alternatively, when no ring beam is used, the horizontal thrust from the sloping RF beams can be absorbed by the stiff connections between the structural members (beam–column connections) and a suitable rigid cladding material. This principle has been used in practice in the RF designs by Graham Brown, described in detail in Chapter 10.

Two-dimensional, In-plane, RF Structures

There are early examples of plane RF structures used in grids of floor beams. The flat grillages presented in the history part of the book (see Chapter 2), designed by Sebastiano Serlio, Leonardo da Vinci and Villard de Honnecourt, are examples of this type of structure. These planar frames have members arranged very similarly to the frames with sloping beams described above. The unique interlocking arrangement of the members ensures that the structure is stable and acts in a similar manner to that of a moment frame – that is, a frame with stiff, fixed connections that can transfer bending moments.

A modern example, though unbuilt, of this type of planar RF frame is the concrete RF structure of the Mill Creek Housing Project by Louis Kahn, described in the history section of the book (see Chapter 2).

RF Structural Models as Examples

Using GSA structural analysis software by Oasys, examples of different RF structures have been analysed. The following RF structures have been chosen as representing typical examples of both the flat RF structure and the inclined (roof) RF structure:

1. Flat (in-plane) RF structure with four main members and an overall diameter of 7 m and an internal diameter of 3 m.
2. RF structure with members inclined at a relatively steep angle to the horizontal. There are eight main members and the overall diameter of the structure is 7.9 m and the internal diameter is 1.2 m.
3. RF structure with members inclined at a relatively low angle to the horizontal. There are eight main members, the overall diameter of the structure is 7.9 m and the internal diameter is 3.3 m.

For all three models a single load case has been considered with symmetrical vertical load applied at the beam intersection points (as a simplified dead load or self-weight). The total vertical load is the same in all three examples.

The results of the analysis of these three models are shown in Figures 5.1–5.3. The results will be discussed in the following sections.

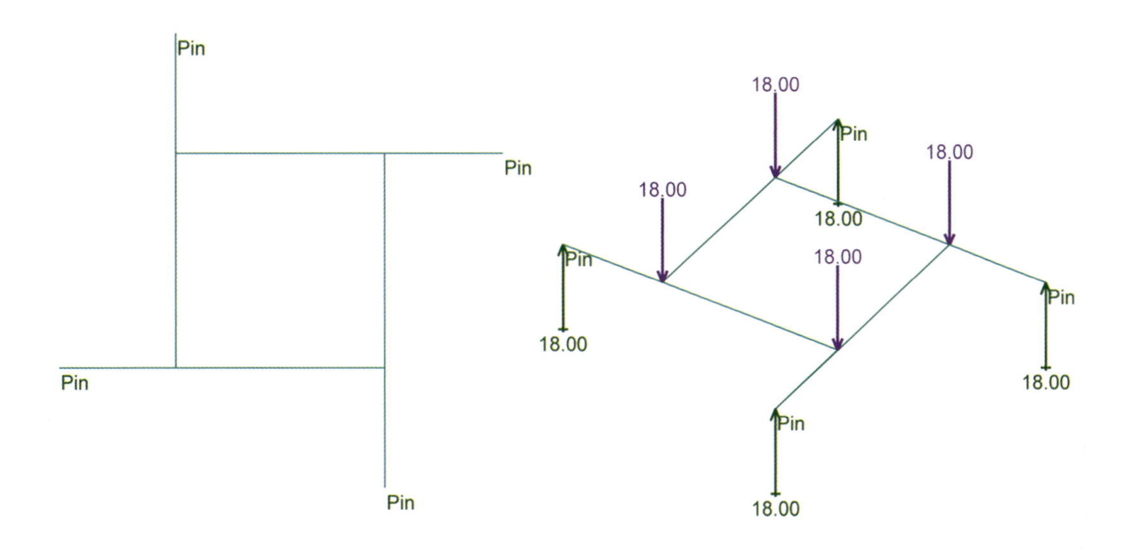

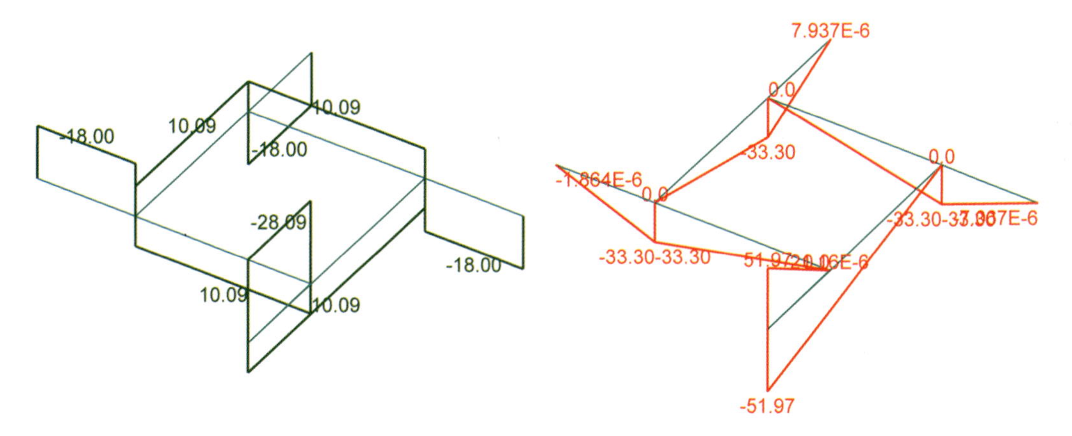

▲ **5.1** Model 1 – flat RF structure with four beams: plan, loads, shear and moment diagrams. (Drawings: Jens Larsen.)

Axial Forces

In RF structures with inclined beams, axial forces are distributed through the members. The lower part of the beam, between the outer support and the point where the beam is supporting the adjacent one, is in compression, whereas tension forces will occur in the upper part of the member between the support at the inner end and the point of support of the adjacent member.

For model 2, with steeply inclined members and a smaller central opening, the compression in the members is nearly twice that of model 3, with low pitch and a large central opening.

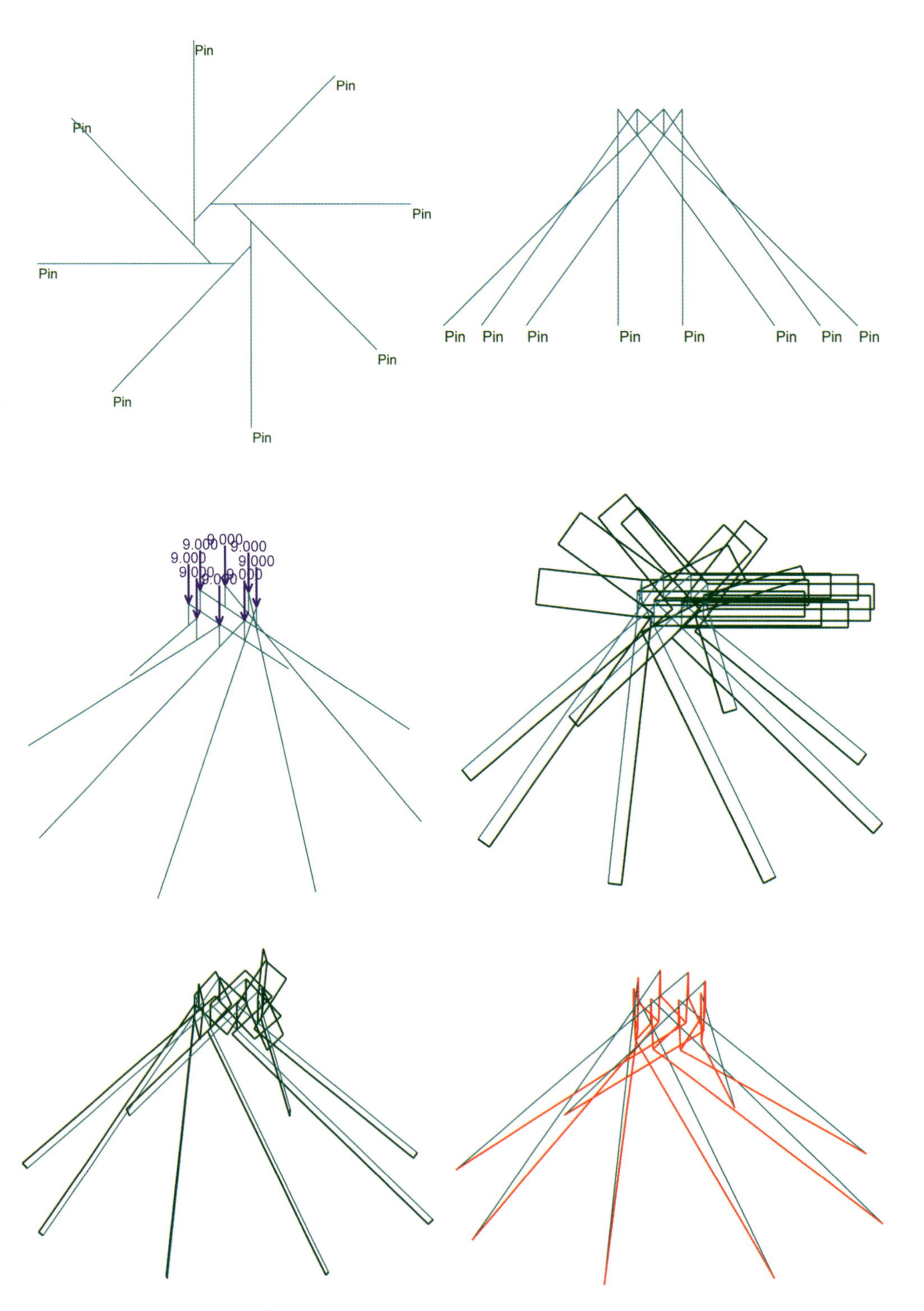

▲ **5.2** Model 2 – steep RF structure with eight beams: plan, side elevation, loads, axial, shear and moment diagrams. (Drawings: Jens Larsen.)

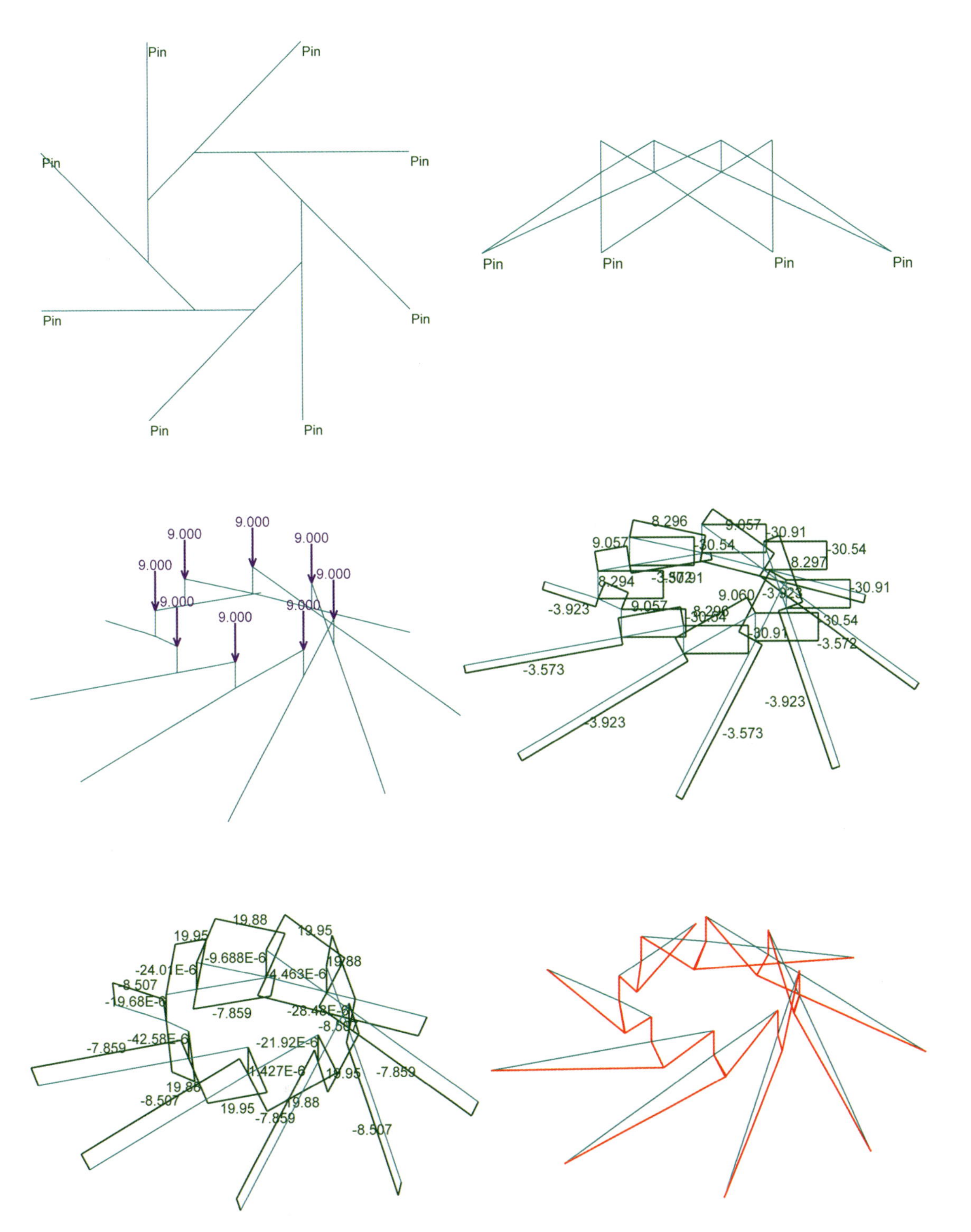

▲ **5.3** Model 3 – low-pitch RF structure with eight beams: plan, side elevation, loads, axial, shear and moment diagrams. (Drawings: Jens Larsen.)

For the planar type of RF structure (model 1), there are no axial forces in the members, because the vertical load is transferred to the supports purely through bending and shear, as shown in Figure 5.1.

Shear Forces

In all three of the above types of RF frame, with sloping beams and planar systems, each beam will deliver a point load at the inner end of the adjacent beam, and this load will create shear forces in the supporting member. The amount of shear force transferred through the beams is related to the slope of the beams and the size of the inner opening. It will increase with the decrease of the inner diameter – that is, the smaller the inner opening, the greater the shear force in the members. This is also clearly illustrated when comparing the shear force diagrams of models 2 and 3. This is important to bear in mind when designing RF structures with timber members, because shear forces can be critical due to the relatively low shear strength of timber.

The issue of the shear forces, which become greater for a smaller inner opening, is reflected in how RF buildings are designed practically. Graham Brown's RF designs (see Chapter 10) usually have relatively small inclinations of the beams and relatively large roof openings, which reduce the magnitude of the shear forces. In addition, in his designs he uses relatively large cut timber or glued laminated members and in some cases, because of the notched detail which weakens the RF beam at the point of high shear forces, he has strengthened the connections with metal connectors.

In the New Farmhouse RF building designed by Yoichi Kan (see Chapter 8), the inner opening of the roof is very small and the roof slope is quite significant. Although the shear forces are considerably higher than if a design with a larger opening and flatter roof had been used, because the designer avoided the notched connection, which would have weakened the beams, he was able to produce a design with very slender beams with a depth/span ratio of 33. This was also possible due to the way the roof was configured. Normally, over a square floor plan, one would use a four-RF-beam design. However, in this particular design, where eight RF members are used to form the main roof structure, the overall roof load is shared between eight beams instead of four, which also made it possible to use very slender beams.

Bending Moments

The inherent geometry of the reciprocal frame means that the occurrence of bending moments in the main members is unavoidable. The point of support of the adjacent beam or rafter member incurs a peak moment, which, together with the shear force at the same location, will determine the necessary size of the main members for a given geometry.

Generally, we would expect that the bending moment would increase with the size of the inner diameter, because the point load from the adjacent member moves further towards the middle of the beam. However, when comparing models 2 and 3, the moment in model 2 is marginally larger than in model 3, although the inner diameter is considerably larger in model 3 than in model 2. This is presumably due to both the effect of the different roof pitch and the fact that the load is moved further towards the centre in model 2, thereby giving overall larger forces in the structure.

Geometry

The relationship between the geometry of any given RF frame and the internal forces is complex. The size of the forces will be dependent on the following parameters:

- The outer diameter or overall span of the structure
- The inner diameter or opening
- The pitch of the structure
- The depth of the main members
- The number of main members.

As a general rule, the larger the inner diameter, the flatter the RF. If a relatively small central opening is required, it will be necessary to have relatively steeply inclined members. This is due to the need for each member to touch, such as the RF members in the design of the New Farmhouse by Yoichi Kan (presented in Chapter 8), or be notched into its adjacent supporting member. Most RF designs, including the RF designs of Graham Brown (presented in Chapter 10), as well as the RF structures designed by Kazuhiro Ishii (Chapter 7), use notched connections. On the other hand, if a connecting vertical piece can be inserted at the interface between the adjacent main members, it becomes possible to have both a small central opening and a relatively low-pitched structure. However, this would not be an RF structure, as defined, with beams touching and supporting each other.

Refer also to Chapter 4 on geometry.

Loading

Apart from the structure's own weight, any RF frame may be exposed to a number of other loads, such as any of the following:

- Vertical load from roof materials, ceilings or other imposed loads
- Snow loads

- Wind loads
- Seismic loads
- Others.

Any of these loads may be asymmetrical and as such will give rise to secondary structural effects, such as uplift, torsion, bending and so on. These effects have to be taken into account in the design of the structure.

These loads may increase axial loads, shear forces and bending moments in the main members and, in critical load combinations, will determine the sizes of the members.

Materials

RF structures can, in theory, be constructed of all the main construction materials (steel, timber and concrete). However, the complex geometry of three-dimensional RF structures, and the need to keep the self-weight low for practical reasons, means that (precast) concrete is not normally a preferred material. For smaller structures, from 2 to 3 m up to approximately 12 m overall span, timber will normally be the preferred material. If the designer has a clear understanding of the RF geometry, timber members can easily be pre-cut and brought ready for construction to site. For steel, the connections between the main members will be potentially complicated to design and fabricate.

Most RF buildings built to date are in the 3–12 m range of span. It is not surprising, therefore, that most of them are constructed from ordinary or glued laminated timber. The only built example in steel known to the author to date is Ishii's Spinning House in Tokyo, described in Chapter 7. As far as the author is aware, the only RF design in concrete is the Mill Creek housing project by Louis Kahn, described in the history section (see Chapter 2), which unfortunately was never built.

Connections

The design of the connections between the RF beams, as well as beam–column or RF members to the ring beam, is important for the behaviour of the structure. The connection design is also crucial for achieving ease of fabrication and erection.

The connections between RF beams when constructed in timber can be achieved by notching one beam and fixing the other into the notch. The notch weakens the RF beams at a critical point where the shear forces are high. The notch is quite complicated and has to be designed very carefully. The timber RF rafters have to be pre-cut very skilfully. If high precision is not achieved, the roof will not fit together. However, this type of connection creates a certain architectural expression. The interlocking beams are interlaced between each other. Most RF buildings built to date use this type of connection.

▲ **5.4** Notched connection – Graham Brown's RF building under construction.

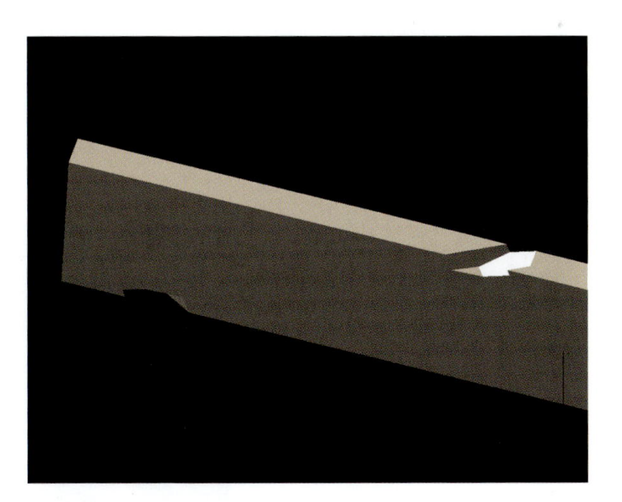

▲ **5.5** Notched RF beam.

Another way of connecting the RF beams is to use friction and to place them on top of each other, which is structurally more efficient, as described earlier. Kan's design of the New Farmhouse building is the only one that uses this type of connection (see Chapter 8).

▲ **5.6** Friction connection – the New Farmhouse RF building by Yoichi Kan under construction. (Photo: Yoichi Kan.)

A third option would be to use a pinned connection and to build up the RF members where they touch each other. This would be structurally very efficient as it would not decrease the depth of the section of the members at the point of highest shear. Also, this type of connection would be easier to make as there would be no need for pre-cutting complicated notches. However, to the best of the author's knowledge, this type of connection has only been used on small-scale models and no full-size buildings have been constructed to date.

▲ **5.7** Built-up connection – physical model.

▲ 5.8 Built-up connection - detail.

All the described connections are for joining together rectangular timber RF members. For round-wood sections a tied bamboo type of connection can be used, or the RF members can be drilled through and connected using metal ties. Examples of this type of connection are used in the design of the Roundhouse, presented in Figure 11.4 (Chapter 11) and also in Graham Brown's Earth sanctuary presented in Figure 10.12 (Chapter 10).

When building RFs with other materials such as steel, concrete or others, the connections will need to be designed to be appropriate to the material used.

Forming the Roof

The architectural expression of RF structures differs in two basic ways: firstly, in roofs where the RF is observed both internally and externally; and secondly, where it is only expressed internally, i.e. one only becomes aware of the RF structure upon entering the building.

All Japanese examples, with the exception of Kan's New Farmhouse RF structure, are designed so that the RF structure becomes apparent when entering the building. The tiled conical roof surfaces that enclose the roof conceal the RF and are supported by a secondary structure. The unexpected structural form of spiralling beams, which only becomes apparent inside the buildings, surprises the visitor to some degree.

The other way of forming the roof structure is through panels which span between the RF beams. If the panels are fixed on top of one beam and attached to the side of the next, the external form of the roof comes to resemble a turbine. All RF buildings designed by Graham Brown (see Chapter 10) use this type of roof. In most cases his buildings have used timber shingles for cladding the roof. Only in the case of the Findhorn whisky barrel houses was copper cladding used as a finish, making the lightweight roof appear heavier than it actually was. Enclosing the roof with panels forms a specific type of architectural expression, which one may or may not like. It also takes away the element of surprise because the RF structure is evident both externally and internally. Depending on the size and proportions of the buildings, it works better in some designs than others. The small gazebo buildings designed in this way by Graham Brown are good examples, having proportions that work well. However, probably the most successful example is the RF chapel roof at Colney Wood, described in detail in Chapter 10. For Graham Brown this is the only 'right' way to design RF roof structures.

Obviously, these are not the only ways in which the RF roof can be formed. Membranes or other structures can be supported from the main RF structure, which would create new and unexpected forms. Although there are as yet no built examples constructed using these alternative roofing solutions, there is no reason why they could not be built.

Progressive Collapse

Progressive, or disproportionate, collapse is an inherent issue with RF structures. The structures rely on interlocking of the main members, which means that the accidental removal of one member can potentially mean the collapse of the entire structure. The building regulations, codes of practices or national standards in many countries stipulate that the risk of proportionate collapse must be addressed in the design. Normally, the regulations will specify what type of buildings or structures, or in what proportion of the structure, collapse would be acceptable if a single member was accidentally removed. Therefore, for relatively small RF structures, or for lower risk building groups (for example, low-rise dwellings or agricultural buildings), the design may not be subjected to any special restrictions with regard to disproportionate collapse.

For larger structures or structures of greater importance, where the consequences of collapse are more severe, design measures would need to be included to deal with the risk of progressive collapse. These may include additional tying of members that act as diaphragms and allow individual members to be held up by catinery action. Also, in seismically active areas it would be important to ensure greater structural redundancy. In the case of the Japanese RF structures, the earthquake energy is dissipated by designing timber joints without steel connectors, which allow the structure to move with the earthquake motion. An excellent example of increased structural redundancy is the Seiwa Burnaku Puppet Theatre Exhibition Hall RF structure, where a double RF structure of clockwise and anticlockwise spiralling RF beams is used for the roof. In the case of the progressive collapse of one of the RF structures, the other will take over the redistributed load of the roof. Further research needs to be carried out to explore the structural behaviour of RF structures, especially when they are subjected to dynamic loadings. Despite the fact that there is a need for further research, it is clear from the Japanese built examples that progressive collapse is a problem that can be successfully resolved.

6 JAPAN – A HOME OF RF STRUCTURES

The next three chapters look in detail at the work of architects Kazuhiro Ishii, Yasufumi Kijima and engineer Yoichi Kan, who have designed several buildings using reciprocal frames. Although the work of each of these Japanese designers is quite different, they all have a very strong unifying element – the reciprocal frame. As can be seen from their designs, these designers have used the structure in different ways, and as a result have achieved different types of space and very different kinds of architectural expression. It is interesting that none of them refer to the structure as a 'reciprocal frame' and even more interesting that the inspiration for their designs came from very different sources. However, all the reciprocal frame (RF) designs, although truly contemporary, show great respect for Japanese culture, its tradition of timber construction and its indigenous architectural values.

The designs of Ishii, with the exception of the 'Spinning' house, are all of timber construction and have been influenced by the idea of 'movement spaces' and Sukiya style. The latter is an especially important influence in the design of the Sukiya Yu residence.

Engineer Kan is influenced to a great degree by traditional Japanese timber architecture and through his creation of the Torikabuto Life Sciences Laboratory, he transforms a traditional farmhouse design into the New RF Farmhouse building.

Architect Kijima shows great respect for the masonry craft of the region in his design for the Stonemason Museum. Although the building is, in every sense, a piece of contemporary architecture, influenced to a great degree by Buckminster Fuller's structures, at the same time it celebrates the values of hand crafting of the stone that is specific to the region.

It is very difficult to compare the Japanese RF buildings without oversimplifying their design ideas and values. However, they all have in common the juxtaposition of the old and the new, and it is this weaving of tradition and history into the contemporary that makes them so special.

The RF is used as part of the designers' idea; it complements the main values of the architecture it is part of.

Before analysing the Japanese RF buildings in some detail, it is perhaps important to say something about the use of timber, the tradition of 'movement spaces' in traditional architecture and the characteristics of 'Sukiya', the tea ceremony. In a way, they form the context in which the RF buildings in Japan have emerged.

Use of Timber

Japan is renowned for the use of timber in construction. Throughout Japanese tradition, trees were objects of worship and the 'godly nature of trees has been raised to an art which can be felt in the architecture of wood' (*Process – Architecture*, 1981). The main use of timber construction was a 'post and beam' structure, most probably as a protection against earthquake. Timber construction in Japan has been developed to perfection, especially in the details of timber joints, which are a very sophisticated method of dissipating earthquake energy. Wood has always been used with special care, one of the reasons probably being religious. There were beliefs that when a timber temple is destroyed by fire, the spirits of the trees used in the building ascend to heaven. Timber was a 'living' thing, therefore when used in construction it was always installed in the structure in the direction it grew, having the root end down. Japan is probably the only country in the world where timber is stacked standing as opposed to the conventional horizontal method in most Western countries.

Most Japanese temples, houses, prefectures and other traditional buildings have been built from wood. It is not surprising, therefore, that the largest traditional timber building in the world is in Japan. The Todaiji Temple in Nara is 57 m wide, 50 m deep and 47 m high, and houses the Diabutsu, or Great Statue of Buddha. The building dates from 1708 and is only two-thirds the size of the original, which had been destroyed by fire (Chilton, 1995).

The Concept of 'Movement' Spaces in Japanese Architecture

When one looks at seventeenth to eighteenth century Japanese spaces and planning and compares them with Chinese examples from the same period, one of the most significant differences is the plan layout. The Chinese have a geometrical organization of the buildings (and spaces) based on an orthogonal coordinate system. Every building and space in Chinese layouts from this period is related to the reference axes, and the compositional elements of the space have to be observed simultaneously.

They were considered a good piece of design if they formed a 'prospect' or a 'vista'.

On the other hand, the Japanese buildings and spaces of the same period are mainly characterized by asymmetry, irregularity and indefinite organization. There are no axes to which all spaces are related: only the preceding and the proceeding spaces matter. A new scene is discovered at every turn and left behind at the next space. The emphasis is on the relative positions of spaces and rooms, rather than axes. Inoue (1985) refers to these types of space as 'movement' spaces, as opposed to the 'geometrical', in the case of the Chinese temples. Figure 6.1 shows two diagrams of such 'movement' spaces. When one is in space 'D' there is an awareness of the existence only of spaces 'C' and 'E', and as one moves through the building one becomes aware of the next approached space. Although the two diagrams seem quite different, there is no significant difference, because the relationships between the internal spaces are the same in both of them.

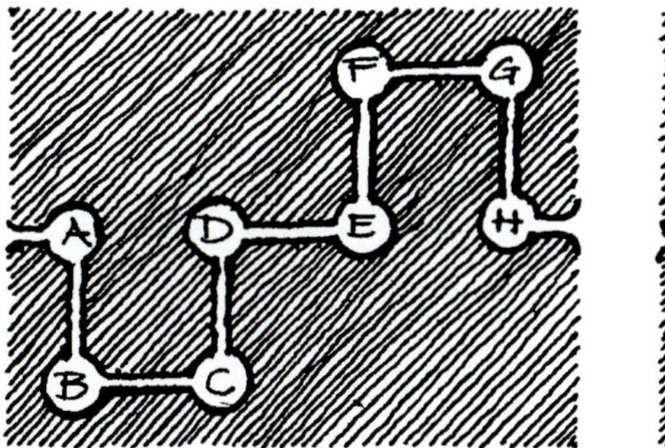

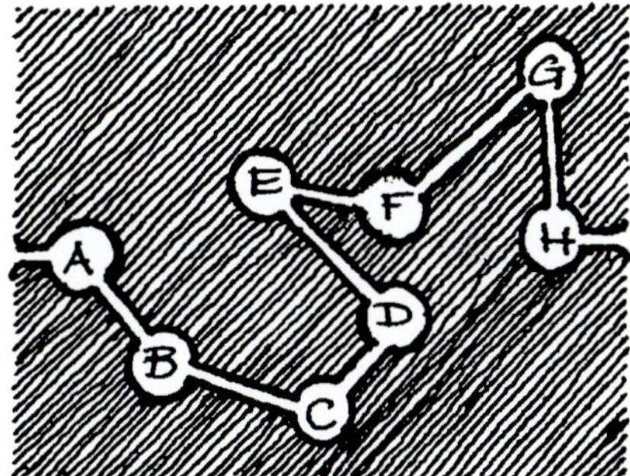

▲ **6.1** Diagrams of 'movement' spaces. (Sketch by A. E. Piroozfar.)

The concept of 'movement' spaces is one of the major characteristics of Japanese traditional architecture. The expression of movement in plan results in zigzag patterns, with buildings and spaces organized in a 'U' or diagonal layout. The fragmentation of spaces in plan was a major contributory factor in the creation of the concept of 'movement' spaces. Most of the elements of each building are designed so that they aid the formation of the 'dynamic' composition.

The layout of some Japanese towns also suggests movement. While the Imperial capitals, 'miyako', were very much influenced by Chinese town layout (symmetrical) (Masuda, 1970), the castle towns followed in great detail the 'movement' concept and had irregular town planning. Most modern Japanese towns have been influenced by them and kept their irregularity. If one questions the reasons for the development of this type of layout, one of the most likely explanations would be the creation of 'defensive spaces'. The Japanese people have been recognized throughout history as excellent warriors. In order to confuse their enemies they designed town layouts with spaces which were not easy to move through.

It is very important to emphasize that movement did not occur only in plan. Traditional Japanese buildings gave a sense of three-dimensional movement as well.

As in upward spiralling, 'movement' in the vertical direction is expressed very strongly by the way in which the castle roofs are arranged. Although the alteration of the roofs as they progress vertically is irregular, it suggests a spiral composition and rotating movement. The roofs of Nagasaki Castle presented in Figure 6.2 change from level to level. They are a feature unique to Japan.

▲ 6.2 The roofs of Nagasaki Castle.

The Japanese spiritual idea of mutability has importance for the whole concept of 'movement' spaces. It comes mainly from the Buddhist religion, which looks at all living things through their 'flowing movement through the three worlds of past, present and future' (Inoue, 1985).

Bearing all this in mind, it is not at all difficult to visualize the RF concept in the context of Japanese traditional architecture. The structure itself suggests movement in the vertical axis. The beams which support each other give the notion of frozen upward spiral motion.

The 'Sukiya' Concept

The Sukiya is a style used mainly for residential architecture and was developed in the fifteenth century. As Itoh (1972) stated, 'it evokes a world of associations with buildings in which the traditional fondness for natural materials, simplicity, and closeness to nature dominates every detail of the composition'.

Sukiya developed as a result of consideration of the aesthetics of a house, the search for its own beauty, so that the 'sakui' (the creative will) of the individual had the highest priority. The Sukiya concept is very important because it stresses the importance of individualism and creativity of design for the first time in Japanese architectural history.

The word 'Sukiya' means tea house in its basic sense, but in its broadest sense it is any structure built with the architectural techniques of the tea house (Itoh, 1969).

Kazuhiro Ishii (1978) described the essence of the tea house as:

> '... *a coded image of habitation which can be regarded as being connected to a return to the womb as a primordial mode of existence. In this sense the tea house is ideal – "environmental". In a twilit space you become sentient in the most complete manner. Here a world of relationships unfold, not a world of denial. The sensation of movement, and the senses of hearing, smell, taste, touch, sight and time as well as sexual feeling are all wide awake in your body, seeking communication in an outward embrace. Your sense of hearing will be at its most sensitive to the boiling sound of the tea-water which has been said to strike the chord of an ear listening to the voice of a pine-cone, the sound of the winds whizzing by outside, the subdued rustling of the kimono of those present, the rubbing sound of tabi against the tatami, the sound of sliding paper doors being opened, the sound of hot water being poured from a tea-water dipper, the sound of the handle of the dipper hitting the rim of the iron tea-kettle, the sound of the handle of the bamboo tea-stirrer hitting the teacup, the sound of tea-sipping, the faint sound of breathing, the voices of people speaking, the sound of wiping the tea-ceremony paraphernalia, the sound of symbolic "dotaku" bells, etc. The variety of taste of sweets and the deep, bitter taste of tea in harmonious interaction, the tastes of fishes, mountain plants, shells, meats, sake, etc. served before tea. Then we have the smells of tea, incense and charcoal. The smell of charcoal, faint but distinct, appears to*

> *carry with it a subtle suggestion of warmth. A sense of temperature is assured by the warmth coming from the charcoal burning in the heart, the hot tea and the symbolical warmth of your heart.'*

The 'movement' concept in Sukiya comes from the spatial composition of the tea house, which is layered, and is a complex assemblage of small space units under a single roof. The Sukiya buildings provide us with 'discoveries' as we approach the next space. One never quite knows, while walking through a building, what the next room would be, whether a small or a big space, a banquet hall or a tiny tearoom, or an inner garden instead. All the typical features of the 'movement' concept, such as asymmetry, irregularity and indefinite organization described previously, apply to a great degree to Sukiya buildings.

Looking at the work of Ishii, Kan and Kijima, presented in detail in the next three chapters, it is evident that although these three designers have been influenced in different ways and by different factors, traditional Japanese architecture has had a great role to play. Their designs differ in how the 'old' and the 'new' come together and in how the contemporary influences have been juxtaposed with the traditional.

It is that approach that makes these designs what they are – unique forms of RF architecture.

7 The reciprocal frame architecture of Kazuhiro Ishii

The Initial Meeting

I feel excited and nervous, standing in front of a cake shop at Akasaka tube station in central Tokyo. It is nearly noon on Tuesday 21 November 2006 and I am due to meet Kazuhiro Ishii, the architect who has designed the greatest variety of reciprocal frame (RF) buildings and, in my view, the most beautiful ones.

Ishii arrives spot on time and I recognize him easily, as I have seen photos in the numerous publications about his work. He suggests that we have lunch together and on the way to the café, he shows me a few of his designs in Akasaka. He tells me that this part of Tokyo is changing very rapidly. In the past it was known as the red-light district of Tokyo and was built up with low-rise, low-quality, housing intermingled with small shops and cafés. Now, however, this district is developing into a prime location: the old shops are being transformed into new trendy ones, while the two- or three-storey buildings are being replaced by skyscrapers almost overnight. It is a place of great contrasts.

As we walk, Ishii tells me about his concerns about the pollution of our planet and his strong belief in environmentally responsible design. He says:

> *'We as architects have a role to play and it is our duty to help future generations. At present we use too many man-made materials in construction. As a result we pollute our planet with emissions of gases such as CO_2 and other greenhouse-effect gases. A few years ago the Akasaka local authorities approached me to design the street lighting for the central streets of Akasaka. Straight away I felt I could express my views about the importance of using more timber in construction instead of man-made materials by using timber in my design. Timber has been used for hundreds of years in Japanese architecture. I have always been interested in Japanese history and am strongly connected to Japanese culture and*

traditions. Every place has a history and a life of its own. We as architects must understand that nothing starts with us. We must try to understand the history of the locality for which we are designing. I come from Tokyo and I know Akasaka very well. For this particular design, the street lighting, my starting points were environmental issues and how to make people aware of them, as well as my interpretation and understanding of Akasaka as a locality with its tradition and history. All my designs carry a thread of tradition in them but really they are contemporary. History is used as a starting point, inspiration and translation in my designs, but not a source from which to copy.'

▲ **7.1** Sando-casa – bird's eye view. (Photo: Kazuhiro Ishii.)

▲ **7.2** Sando-casa – this is not a tree.

As we walk, Kazuhiro Ishii stops and points out a street lamppost in front of us that appears to be made of timber, its column a real tree trunk with its bark still on and the fitting at the top resembling a medieval warrior's hat. At first I think that I am deceived because the street light seems not to be vertical, but rather leans away from the street. But as I lift my eyes and look ahead, I see that all the street lights are leaning. I cannot stop myself from touching the tree trunk. I don't expect it to be real, but to my great surprise I find that it is. I like the feeling of the rough bark of the tree on my hands. I ask Ishii to tell me the story of the tree trunk street lights, as I have never seen anything like this before. We continue walking through the lively streets of Akasaka as he explains:

> *'I didn't want to miss the opportunity to do more than just design street lights that simply looked nice. I wanted to make people stop and think about the future of our planet. My design for the street lights – I call them Sando-casa – tells the story of how wood should be used more in a symbolic way. I am aware that it cannot replace all man-made materials, but I think that it should be used more often. It is an environmentally friendly material and its use reduces CO_2 emissions. In the case of these lampposts we had to use a steel insertion into a hollowed timber tree trunk in order to make the lamppost self-supporting, and to bring electrical power to the lamp at the top. It is a kind of a local oak tree known here as "kunugi". So, you could say the message is symbolic, because I still use steel for the lampposts. It is not a timber-only lamppost. Still, I feel the narrative of my design is intended as a reminder to people about our responsibility towards nature and with an aim of reducing the pollution of our planet.'*

▲ **7.3** Sando-casa – the leaning lampposts. (Photo: Kazuhiro Ishii.)

▲ **7.4** Steel core reinforcement. (Photo: Kazuhiro Ishii.)

I then ask about the unusual fitting at the top: 'Why a hat shape? Also, why are the lampposts leaning?' Ishii continues by asking me a question:

> *'Have you ever seen a perfectly vertical tree? You would probably not be surprised if I told you that there are not many around. As with the lampshade shape, my inspiration is in Japanese history. We are very proud of our brave warriors – "Yakuzas" – who lived in the past. The form of the lampshade is inspired by the form of a traditional Yakuza travelling hat. It is not a copy of it, but a Japanese person would recognize the resemblance. As you can see, history and tradition are very important to me. Our knowledge and understanding of our past and our culture makes us what we are.'*

▲ **7.5** Lampposts ready for installation. (Photo: Kazuhiro Ishii.)

I then comment that planning authorities in Japan must be very forward thinking to allow these unusual-looking lampposts to be constructed. Ishii's laughter tells me everything. He explains:

> *'Nothing in life is easy, but I never give up, I question, I negotiate. I fight for what I believe is right. As an architect, I feel it is my duty to change people's views for the better. Local authorities in Japan are very conservative and it took some convincing to get planning consent for the Akasaka street lights. I wanted them to be leaning more at first, but we agreed that 3% of the vertical is sufficient to make it obvious that they are leaning. In the beginning the planners did not want to know. They were so opposed to the whole idea of the leaning lampposts that they did not want to hear about it. But in my*

> *negotiations with them I explained that Yakuza's travelling hats were always tilted (by 3%?![1]) and that was another reason for having the tilt. At that point they gave up. But it took some time and some persuasion. It was fun.'*

It is not surprising that the cost of constructing these lampposts was considerably higher than that for any conventional lampposts. Tree trunks that were relatively straight and of a particular width had to be transported from Ibaraki prefecture; a new machine for hollowing the tree trunks had to be constructed to speed up the process of insertion of the electric cables and steel columns, all of which contributed to the high cost of the lampposts. However, Ishii proudly tells me that the clients (all the shopkeepers) were so pleased with his design that they were happy for the cost to be higher. It is interesting that some time later I find out that Ishii's design for Sando-casa (the leaning lampposts) has been awarded prizes by the Ministries of the Environment; of the Economy, Trade and Industry; the Ministry of Land and Transport, as well as by the Forestry Agency of Japan. Ishii did not mention any of these prestigious awards when he was talking about his design ideas for Sando-casa, and it was months later that I found out about them.

▲ **7.6** Sando-casa – night view. (Photo: Kazuhiro Ishii.)

This is how my initial meeting with Ishii and our short walk to the café went. He is a man of great architectural talent and someone who truly enjoys every aspect of his profession. Also, he is a man with strong views and beliefs. His designs can be understood at a number of levels

[1] I am sure that Ishii is right in saying that the hats of these brave warriors really were tilted, but aren't most hats tilted? How can one measure the angle of tilt of a hat? I think that most people who know him would agree that Ishii has great negotiating skills!

and there always seems to be deeper meaning in them. As we walked together I realized that he is able to explain the philosophical depth of his designs with an unexpected simplicity while remaining very approachable, modest and easy to talk to.

Kazuhiro Ishii graduated from the School of Architecture at the University of Tokyo in 1967, where he studied under Arata Isozaki. Between 1972 and 1975 he studied at Yale University under Charles Moore and James Stirling. After his return to Japan in 1976, he set up his own practice. He has lectured at Waseda University, Tokyo, the University of California, Los Angeles and at Yale. His best known works include: A House of Fifty-four Windows, Naoshima Junior High School, Takahashi Residence, Takebe Kindergarten (54 Roofs), the 'Sunrise' and 'Moon-rabbit' villas, the 'Spinning' house, the 'Bi-costal' house, A House of Our Generation, the Puppet Theatre in Seiwa and the Sukiya Yu house. He has also published several books including: *Thoughts on Sukiya, International Architectural Parts* and *My Day at Yale*. In addition, his work has been featured in several TV series as well as in a great number of journal articles explaining the philosophical and cultural background of his designs.

For someone of such high standing in Japanese and world architecture, it is amazing that everything so far makes me feel at ease in his presence.

▲ 7.7 Architect Kazuhiro Ishii.

Although by this time I had only spoken to Ishii for about 15 minutes, I feel already that I know him well. All my nervousness had by now disappeared.

At this point we arrive at the Maroon café. It is a small but very cosy place. There are only four or five tables. The owner, Mr Kurihara, a close friend of Kazuhiro Ishii, quickly makes sure we are comfortable and brings us traditional Japanese vegetable soup. I cannot wait to hear all about Ishii's reciprocal frame designs. I wonder whether what I had read about them is how they were really conceived. What had been the architect's inspiration?

During the meal at Maroon café, a traditional one consisting of many beautifully arranged small dishes, Ishii tells me about his reciprocal frame architectural designs and about other buildings too. The conversation is spontaneous and we only get interrupted when another small dish is brought to the table by the kind café owner. We talk first about the Enomoto residence, the 'Spinning' house in Tokyo.

▲ **7.8** Human reciprocal frame. (Sketch by A. E. Piroozfar.)

The 'Spinning' House (Enomoto Residence) in Tokyo

The 'Spinning' house was designed in 1985 by Ishii for the Enomoto family. It is situated in the Tamagawa Gauken residential district in Tokyo. It is a steel-framed house with spiralling steel Vierendeel trusses, externally clad with exposed prefabricated concrete panels. The house is located on a small hill in a tight urban site. It is organized over three levels, with bedrooms radially arranged on the ground floor around a central hall. The living room area is on the second floor and there is a study on the third floor. The longest span is about 5 metres. The steel RF structure, made of Vierendeel trusses, is the only part of the building that can be seen from a distance. As one comes very close, the rest of the house becomes visible too.

▲ **7.9** Drawing of an Islamic pattern. (Sketch by A. E. Piroozfar.)

Ishii tells me that the client wanted a different and exciting house, one that would have a lot of light inside the building. Ishii came up with the idea of using an RF structure. Inspired by the method of holding hands, where there is no support for the load at the cross points of an arm and a hand, support being given at the outer end, by Islamic drawings as well as by the spinning of the planets in the cosmos, Ishii created this unusual house.

The 'spinning' effect is achieved by rotating each steel Vierendeel RF truss 15 degrees in relation to the one before it. The effect achieved is very similar to a pop-up tissue box. Ishii states that spinning (whirling) can be found in Islam as a very early expression of the image of the cosmos. Also, the 'movement' concept is very much present in Japanese traditional architecture. The materials and technology used are very

▲ **7.10** Spinning house – external view. (Photo: Kazuhiro Ishii.)

▲ **7.11** Spinning house – night view. (Photo: Kazuhiro Ishii.)

modern. The RF structure contributes to an achieved sense of spinning motion, brightness and light coming from the roof, a sense of floating and refined touch. As Ishii states: 'The roof light formed with the RF makes someone looking feel almost as if they have had a glimpse of the cosmos itself.'

▲ **7.12** Spinning house – close-up. (Photo: Kazuhiro Ishii.)

▲ **7.13** Spinning house – plan. (Sketch by A. E. Piroozfar.)

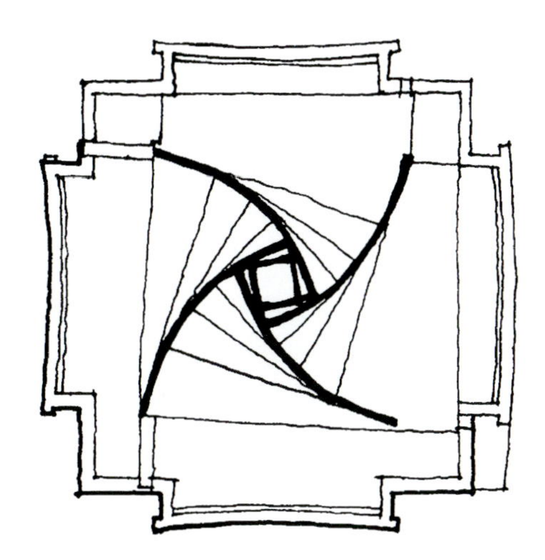

▲ **7.14** Spinning house – roof plan. (Sketch by A. E. Piroozfar.)

There is a lot written and published about the 'Spinning' house. But I want to hear about the design philosophy from the designer himself. So, I ask Ishii to tell me more about this unusual design. He explains:

> *'When you go to a hairdresser you explain what you would like to look like when your haircut is finished. At this point you leave everything to the hairdresser, who is probably someone you trust, which is why you have chosen them. They use their skills and knowledge about fashion to create a hairstyle that they believe will suit you best. A similar thing happens when you ask an architect to design your house. The client for the Enomoto residence told me that they wanted to have more [than usual] light in their living spaces. The rest was left to me. I created a house which may seem strange and unusual. This house was supposed to become a home for my client. You could say that the client was expected to inhabit my creation, which was based on my design ideology. Some people see the "Spinning" house as the "surprise house", one that brings sensation, that creates the spaces by using unusual and bizarre elements and a feeling of movement.'*

▲ **7.15** Spinning house – interior view. (Photo: Kazuhiro Ishii.)

▲ **7.16** Spinning house – interior roof view. (Photo: Kazuhiro Ishii.)

Ishii continues:

> *'At times there could be a gap between the client's expectation and the architect's ability to translate it into a piece of beautiful architecture based on their design ideology. In extreme cases the client could feel trapped into the design ideology of the architect, just like a prisoner is locked in a cell. I always question the role of the client and the relationship with the architect. In my view the client has a great role to play with their input into the design. However, it is important that they [the clients] are open-minded and prepared to expand their views. In most cases I have been lucky to work with open-minded clients who have trusted my skills and have been able to trust me. Of course, we negotiate[2] and work closely together. In the case of the Spinning house, I was inspired by the universe and the rotation of the planets. In the time before Galileo, humans believed that all planets spin around the earth. Galileo freed us from the religious dogmas and made us aware that the planets spin around the sun. You could say that he took away our self-centred and false view, and also that he liberated us by making us understand our real position in the universe. Through that understanding we are made to feel part of the spinning universe and part of modern society. All these ideas are ingrained in my design of the "Spinning" house. Again, I am pleased to say that the client was happy with my ideas and accepted them. The house I had designed became a home for the client in the way I could only hope for.'*

Although the house has weathered and aged over the last 23 years it still looks striking, with its spinning roof structure that seems to bring the universe into the house.

▲ **7.17** Structural engineer Tadashi Hamauzu has engineered all Ishii's RF designs. (Photo: Hamauzu.)

Sukiya Yu House – Ishii's Reciprocal Frame Design Creates a New Contemporary Sukiya Style

Despite the fact that Kazuhiro Ishii is extremely busy, I am very pleased that over the remaining few days of my stay in Tokyo he agrees to meet and talk about his work several times more. He kindly arranges for me to meet his client Mrs Yasuda, the owner of the Sukiya Yu house in Okayama Prefecture, as well as Tadashi Hamauzu, Ishii's structural engineer in his engineering consultancy in Tokyo. Mr Hamauzu has done the structural engineering design for all of Ishii's RF projects and has worked with him for over 25 years.

The next building we talk about is the Sukiya Yu house, where the RF structure creates the roof of the guest-entertaining building. The house was built in 1990 in Asakuchi-gun, in Okayama Prefecture. The client was

[2] I am sure Ishii's negotiating skills have some role to play!

Mrs Yasuda. She had seen the cultural centre, school and swimming pool buildings that Ishii designed at Naoshima Island and was impressed by his work. She approached him and commissioned him to design her retirement home. It should be mentioned that Mrs Yasuda was a wealthy client. Her husband and son were running the family business, Yasuda Precision, a factory that designs and makes machines for the textile industry, which has a great reputation and exports its products all over the world. Mrs Yasuda had particular views on what her retirement home should look and feel like. She approached Ishii because of his proven ability to create a particular feel and refinement. There was a long process of negotiation between her and Ishii. The design of the Sukiya Yu house took nearly two and a half years and the building took another year and a half to construct. Over this time the design was changed more than ten times and it was only because of the mutual understanding, respect, trust, and the patience of the architect and client that the house was built to the satisfaction of all. As Ishii states:

> *'Without enlightened clients such as Mrs Yasuda, we would not be able to move architecture forward. She was an amazing client. I would not say easy, but she was someone you could talk to, someone with clear views and expectations, but at the same time very open-minded.'*

▲ **7.18** Naoshima swimming pool. (Photo: Kazuhiro Ishii.)

After the house was completed in 1990, Mrs Yasuda moved in and was very happy with the creation. Unfortunately, she only lived for three and a half years in the house before she passed away. Her son and his wife inherited Sukiya Yu and are the present owners of the house. It is the present owner, Mrs Yasuda, that I met and talked to about this amazing house.

▲ **7.19** The complex of buildings forming Sukiya Yu house. (Photo: Kazuhiro Ishii.)

▲ **7.20** View to Sukiya Yu showing the RF guest parlour. (Photo: Kazuhiro Ishii.)

The house is positioned on a relatively big plot of land in beautifully landscaped gardens in a residential area in the small town of Asakuchi-gun, in Okayama Prefecture, with a population of 5000–6000. It is a particularly generous site for Japanese conditions, where houses are built close together and hardly have any garden. Sukiya Yu's entrance is on the densely populated side of the residential district. At the back the house is surrounded by beautiful countryside planted with berries, bamboo shoots and other indigenous plants arranged to complement the design of the house.

Sukiya Yu is an unusual house in that it is not built into one volume as most houses would be. Instead, the house is organized in several small buildings, some of which are interconnected with corridors and some of which are free standing. It forms a small hamlet, a village consisting of several very distinct buildings. Hence, the name of the house, 'Yu', which means 'village consisting of different houses'.[3] I ask Ishii to tell me more about the first part of the name of the house, 'Sukiya', which I am aware is connected to a 400- to 500-year-old traditional Japanese architectural style.

▲ **7.21** Entrance to Sukiya Yu.

He explains:

> *'"Ya", the last part of the word "Sukiya", means house. "Suki" has three meanings: to be fond of; rare (the spelling is slightly different: "suuki"); and transparent.'*

'Sukiya' has been used in residential Japanese architecture and is closely connected to the sensual and spiritual experiences of tea ceremony houses. In architectural terms, Sukiya is a calm and refined style used in the past by wealthy people in Japan for building their residences. The spaces in Sukiya are usually organized as a number of separate spaces attached to a central space. They create a community of their

---

[3] There is another type of village known under the name of 'Son'. Unlike Yu, Son consists of houses which are similar in size, form and architectural style.

▲ **7.22** The sweeping wall marking the grounds of Sukiya Yu.

▲ **7.23** The owner Mrs Yasuda with the small buildings forming Sukiya Yu.

own. The calmness is achieved by refined detailing and use of timber. There is hardly any decoration, especially not any golden decoration. Everything is just calm and refined.

I can see the connection between the old Sukiya style and this house, but at the same time I can see great differences and modern influences. Ishii explains:

> *'After talking to my client, the late Mrs Yasuda, I could understand that she wanted a special and different house, one that could bring calm and refinement. Through the lengthy process of consultation with my client, I realized*

▲ 7.24 The RF building at Sukiya Yu.

that it would be most appropriate to use the old Sukiya as a starting point. In response to my client's wishes and the site, I created a new Sukiya. This new Sukiya is an interpretation of the traditional style, but also has great influences from twentieth century design. I used the refined detailing we can find in old traditional Sukiya, but at the same time I decided to dedicate each building of the Sukiya Yu hamlet to the designers and important influences on Japanese modern architecture that have influenced its present form. So among the buildings of the hamlet you will notice a Buckminster Fuller dome, a Bruno Tout building and an over-exaggerated traditional Japanese temple roof. In addition, I include the influences of four Japanese architects – Taniguchi, Horigushi, Hiroguchi and Tamagushi – who, in my view, have been very important for the creation of this new Sukiya style, represented through my design of the Sukiya Yu house. As through my other designs, I used symbolism to express a message. With this design I wanted to tell people in a symbolic way why Japanese contemporary architecture has developed in a particular way and has become what it is today'.

At this point I ask Ishii about the client. Was he able to explain the philosophical and symbolic grounding of his design? Could the client understand and appreciate the depth of meaning and importance of this design? To this Ishii replied:

'At first the late Mrs Yasuda did not understand the significance and symbolic meaning of my design, but she believed in me, she had trust, and with time she was able to appreciate not only the beauty of the house at its face value, but also the symbolic meaning of this design. Again, I feel I have been

▲ **7.25** Interior view of the RF building at Sukiya Yu.

▲ **7.26** View towards the external sliding doors that bring the garden into the space of the RF building at Sukiya Yu.

extremely fortunate in having an enlightened client like Mrs Yasuda. Without clients like her, all houses would end up looking the same.'

The building in which the RF structure is used is an entertaining space, 7 metres in span, named Yu-an. The horizontally overlapping timber RF beams support the wooden dome. The circular plan, the door openings, the interior with the folding shrines and the construction details are all traditional. With the addition of the wooden geodesic dome, the building becomes an interesting combination of old and new.

The present Mrs Yasuda, as enlightened and cultured as the late Mrs Yasuda seems to have been, is very kind, and spends several hours talking to me and showing me her house. We spend most of the time in the RF entertaining building. It is a free-standing building positioned away from the main house overlooking the beautiful garden. The main space of the building has a roof structure in the form of a Buckminster Fuller dome, which is held up by a double RF structure spiralling in opposite directions. The whole building is enclosed by sliding panels that are formed in the traditional Japanese way. They have three planes: two glass

▲ 7.27 View towards the 'snow' sliding doors.

▲ 7.28 The double spiralling RF structure supports a geodesic dome.

▲ 7.29 The geodesic dome forms the roof enclosure – internal view.

panels, an external and an internal one, and a paper one that can slide in between the glass panels. Mrs Yasuda explains that by this traditional way of using sliding planes to create the external wall panels they are able to control the level of light in the spaces and views to the outside. Very proudly she shows me the 'Yakumi window', the 'snow window', which is formed by inserting paper within the upper part of the glass panels, whilst the lower part is a see-through window. 'By using this sliding arrangement on winter days, when we sit on the floor we can enjoy the snow views towards the garden without losing too much of the internal heat through the wall,' Mrs Yasuda explains.

I stand and look around. I can feel the Sukiya influence in the space. There is calm and refinement. It makes the space yours, owned. You are part of it. I can almost imagine myself being part of a tea ceremony at this very moment in this space. Apart from a screen with Japanese writing there is no decoration in this space. The beauty of the space and its refinement come as a result of the proportions used and the detailing. The interlocking RF beams, beautifully arranged in the double RF spiral, are part of the whole expression. I even think that they challenge and question the old Sukiya and contribute to its reinterpretation into a new Sukiya.

▲ **7.30** Complex notched timber connections.

I ask Mrs Yasuda about the construction of the building. How was the house constructed? She explains:

> *'Mr Ishii had designed a complex roof form and although Mizusawa Construction, the contractors for this project, had 80 years experience in wood construction, this was the most complex project they had ever been involved with. To make sure that all the geometry was right they first built 1:5*

models of the interlocking joints. When they were sure that the geometry was correct they scaled up the notched timber interlocking beams and constructed the roof. We agreed to use untreated Canadian pine for the roof because it was cheaper than to build the roof from local timber.'

▲ **7.31** Constructing the physical model of the RF roof in 1:5 scale. (Photo: Kazuhiro Ishii.)

It is worth mentioning that all the timber beams are pre-cut to high precision so that the members slot into each other, just like a 3D puzzle or a Meccano set. They had to fit perfectly to make the whole structure fit together, especially because there are no metal connectors used in any of the joints. They are held in position because they all slot into each other perfectly. The RF structure in this building is in the form of interlocking beams that form a ring which supports the geodesic dome forming the roof. There are two RF structures in this roof: one with RF beams spiralling clockwise, interlocked with another RF structure consisting of beams spiralling anticlockwise. In this building it is clear that the architect and the engineer have worked closely together. The double spiral of RF beams overcomes the risk of progressive collapse. In the event of an earthquake, if one set of spiralling RF beams loses a member the other spiral will take over and provide structural stability. In addition the complex joints, based on traditional Japanese joints with no metal connectors, also help in the event of dynamic loading. They allow for movement so that if there was an earthquake, the whole building would sway and move with it. Thus, the energy is dissipated, making this an earthquake-resistant structure. Apart from being a stable and earthquake-resistant structure, it is also a very beautiful structure, one whose presence enhances the architect's aspiration to create a new, contemporary, Sukiya style.

Mrs Yasuda continues her story:

> *'Mr Ishii came up with the name for this house, "Yu", which comes from "Yu-an", meaning small village. It is a very unusual house consisting of several buildings just like a small village. My husband and I inherited the house and have not been the clients for it. Had I been the client, perhaps I would have chosen a simpler design, but despite that we are very fond of the house and feel very attached to it. Although the house is a complex of half a dozen different buildings forming a village-like assembly, there is a very strong unifying element that makes it all feel like one house. To me it is the values of the traditional Sukiya that have been brought into the design of the layout and the refined detailing. The reinterpreted old Sukiya has been reinstated in a new and contemporary way. Despite that, it has not lost the refinement that the traditional forms have carried from generation to generation for over 400 years.'*

Only after visiting the Sukiya Yu house could I understand how the new and the old come together. Only then did it become clear how Ishii, inspired by the values of the old Sukiya in planning through the use of different elements, each with its own significance, by the refined traditional detailing and twentieth century influences on Japanese architecture, gave the traditional Sukiya a contemporary resonance.

Bunraku Puppet Theatre

The Burnaku Puppet Theatre designed by Kazuhiro Ishii is set in the town of Seiwa in Kumamoto Prefecture, southern Japan. It is set in the landscape surrounded by dramatic high hills which form a backdrop and a natural border to the site. It is a complex of four distinct buildings, each distinct but brought together through the use of a common architectural language. All the buildings use timber for their structure and all of them except the newly built restaurant use some form of RF structure. The structures are very much part of the overall architectural language, and to a great degree contribute in creating its particular architectural expression. The structures used are all different and define each space in a very sophisticated way.

These examples show how RFs can be designed in a way to give a completely unique and different expression, each suitable for the particular building where they are used. Yet they show the designer's great ability in this complex of buildings: to create distinct and different buildings that are unified by common elements.

The complex consists of four free-standing buildings in the landscape: a Puppet Theatre with auditorium; an exhibition hall building; and the shop

▲ 7.32 Seiwa Bunraku Puppet Theatre complex. (Photo: Kazuhiro Ishii.)

and café in a separate building. A recent addition to the complex is the new building that houses the restaurant.

When Ishii was commissioned to design the Puppet Theatre complex, he wanted his design to help in regenerating the local rural communities, which are in decline. He studied the history and the characteristics of the locality. As always, his approach was to understand the regional issues, the culture, traditions, and by respecting the old to create a contemporary reinterpretation in the form of architecture that links the old and the new in a novel way.

Ishii decided to use wood for the Puppet Theatre complex in Seiwa because of his strong views about environmental issues, but also because he wanted to help the local timber industry. He found writings about a Buddhist monk called Chogen who lived in Nara in the twelfth century and who had used a spiral layering of timber to create structures. Inspired by this, Ishii created the RF structure for the exhibition building.

The RF structure over the exhibition hall is perhaps the most impressive of all the RFs on the site. The exhibition hall is a 13-metre-high space which is flooded in light from the windows and the roof light. It is the building which houses the permanent exhibition of puppet masks, puppets and paintings showing scenes from puppet shows. The hall is a relatively small building of only 8 metres span, but the double height as well as the light that floods the space make it feel a lot bigger than it really is.

▲ **7.33** Seiwa exhibition hall.

▲ **7.34** The double height space of the exhibition hall makes the space feel bigger than it really is. (Photo: Kazuhiro Ishii.)

Part of the architectural expression is achieved by using an RF structure for the roof which is left exposed and is visible in the space. The 12 RF beams that form the roof structure are supported by a woven structure which consists of two flat RFs spiralling in opposite directions and supporting each other. The RF structure is only apparent when entering the exhibition hall, because externally the roof is clad with ceramic tiles laid concentrically on rafters. The exposure of the RF only in the interior of the exhibition hall adds to the visitor's astonishment when noticing the roof for the first time after entering the space.

The tall and slender timber columns in the 13-metre-high space are at the limit of the length allowed by Japanese building regulations. To prevent the slender columns from buckling, the woven double spiralling RF structure is repeated in the form of a three-dimensional ring beam at the columns' half span. Only a 'structurally minded person' realizes the utilitarian function of this ring beam, because it fits so well in the

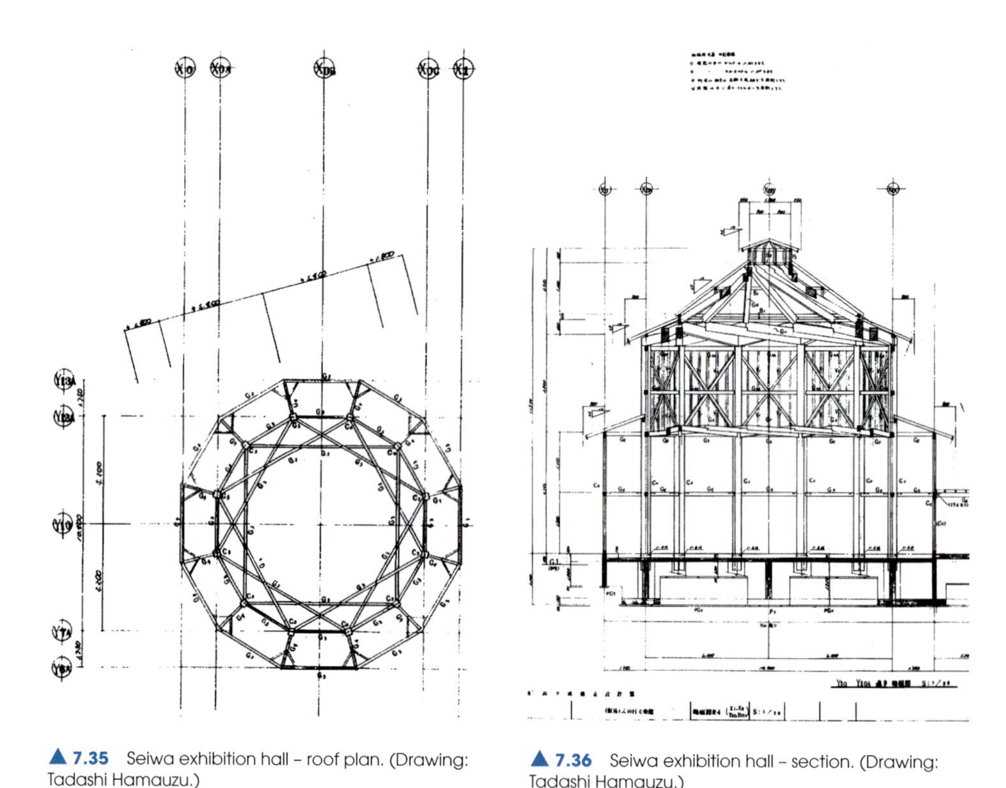

▲ 7.35 Seiwa exhibition hall – roof plan. (Drawing: Tadashi Hamauzu.)

▲ 7.36 Seiwa exhibition hall – section. (Drawing: Tadashi Hamauzu.)

context of the building and gives the impression that it is purely part of the architectural expression. By mirroring and repeating the three-dimensional woven double RF, the space feels more complete. One cannot separate the 'architecture' from the 'engineering' of the building. They are in unity, they complement the 'one' and 'whole' in a way that only a really successful piece of design can.

The detailing in this building is done by using carpentry joints that are based on traditional Japanese 'Vatariago' joints. None of them use any metal connectors. The architect, Ishii, and the engineer, Tadashi Hamauzu, worked very closely to develop the structure that fits and complements the architectural expression envisaged by Ishii. This can be seen by looking closely at the building design. The technical necessities are resolved so that they are part of the architecture. The buckling protection of the columns is clearly part of the overall architectural expression. Also, the ring of RF beams spiralling in opposite directions that support the RF roof structure is relatively heavy, which helps against wind uplift but at the same time mirrors old, traditional Japanese roof structures. It is both utilitarian and beautiful. And, as for the most amazing pieces of

▲ **7.37** The RF-like ring beam reduces buckling.

▲ **7.38** The complex and beautiful RF roof structure of the exhibition hall – internal view. (Photo: Kazuhiro Ishii.)

architecture, it is difficult to decide what came first: the need for a particular architectural expression or the necessity to resolve it in a technically viable way. The two are part of one inseparable whole, a very refined piece of architecture.

▲ **7.39** Assembling the pre-cut RF timber beams. (Photo: Kazuhiro Ishii.)

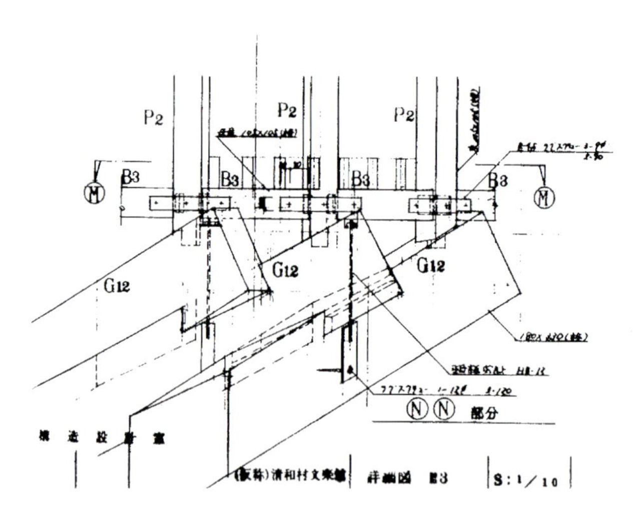

▲ **7.40** Detail of the RF notched beams. (Drawing: Tadashi Hamauzu.)

When I confront Ishii, asking him why he did not use a roof structure consisting of rafters that meet at one point at the top, he simply says:

> *'Look at the universe – it shows a spiralling motion, one that rotates around the centre but avoids it. My roof does the same. This is not a utilitarian building, unlike a castle that in the old days was used for protection and had only one function. There, the beams always used to meet in the centre. An exhibition building is a space that can be interpreted and used in many ways. That is why the structure is one that has a cosmic look, and just like the universe*

▲ **7.41** All RF timber beams are in place – elevation. (Photo: Kazuhiro Ishii.)

▲ **7.42** The structure of the RF roof is in place. (Photo: Kazuhiro Ishii.)

that surrounds us and forms our world, the RF structure in this building creates the "World" of this building.'

To this I can only add: 'Typical Ishii symbolism realized in the most amazing and beautiful way!'

▲ **7.43** Internal view of the RF roof structure. (Photo: Kazuhiro Ishii.)

▲ **7.44** The skeleton of the exhibition hall. (Photo: Kazuhiro Ishii.)

▲ 7.45 Puppets' 'involvement' in the construction process. (Photo: Kazuhiro Ishii.)

▲ 7.46 Puppets exhibited in the finished building.

The other building on the site is the auditorium building, which is connected to the exhibition building via a covered but open walkway. Inside the auditorium building the architect has used a planar grillage structure (referred to in Japan as a 'chopstick structure') to create the roof and the ceiling. One could describe it as a flat type of RF structure, consisting of relatively short timbers that are interlocked and create a woven effect.

The atmosphere and the feeling in the auditorium are very different to the exhibition space. Unlike the exhibition hall that was flooded in light, the auditorium is a very dark and oppressive space. The roof structure,

▲ **7.47** The auditorium building is connected to the exhibition hall via a covered walkway.

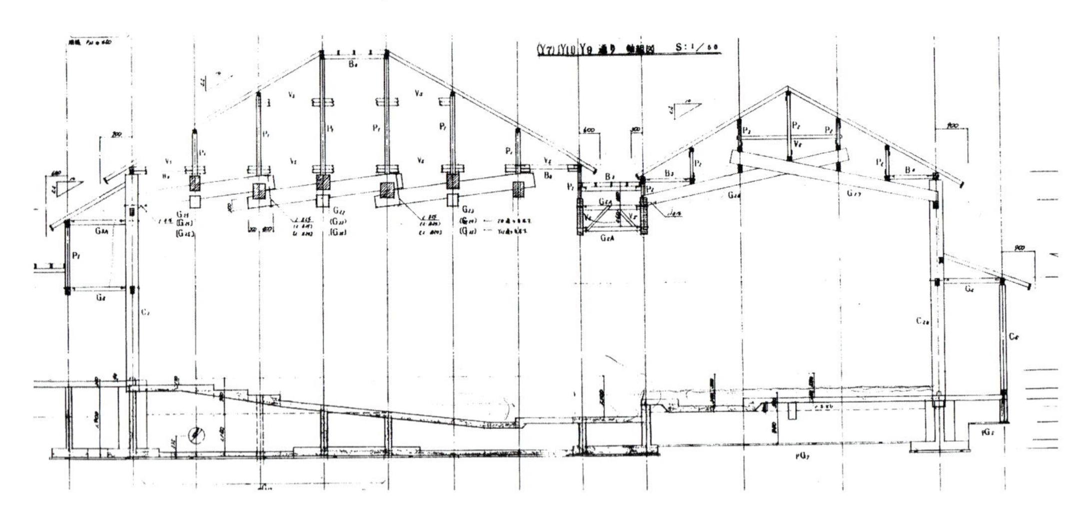

▲ **7.48** Section through the auditorium and stage. (Drawing: Tadashi Hamauzu.)

which is left exposed in the space, adds to the feeling of weight. It is a very heavy interlocked grillage consisting of timber beams that overlap each other to form the roof structure. At the points where the timber beams cross each other and interlock, the overall section (of all three members) exceeds 1 metre in depth. It is a heavy weight hanging over our heads, making us feel the oppressiveness of the space almost physically. I ask Ishii why he used such a heavy structure. He explains:

> *'The puppet stories that are presented in this theatre are of a specific kind. This is a Bunraku Puppet Theatre and the stories are ones that tell us about*

▲ **7.49** Auditorium – internal view towards the stage. (Photo: Kazuhiro Ishii.)

▲ **7.50** Auditorium – internal view of the oppressive space. (Photo: Kazuhiro Ishii.)

the hardship of people who suffered from the Samurai. They are stories about love, money, loss, etc., but they are always sad stories. I felt that it was important to express this feeling of hardship through the architecture of the building. Thus, the heavy timber structure. I based my design on a module used in traditional buildings, "ken". One ken is about 1.8 metres and the roof modules are one ken (1.8 m) or two kens (3.6 m). The whole structure of the roof and therefore the building is designed using this basic module.'

I challenge Ishii with my next question: 'I understand the importance of achieving the appropriate expression for this space by using a heavy

▲ **7.51** Construction of the auditorium roof chopstick structure. (Photo: Kazuhiro Ishii.)

▲ **7.52** Assembling the timber beams' pre-cut beams. (Photo: Kazuhiro Ishii.)

timber structure, but did you not feel that it is wasteful to use up so much timber?'

It is interesting that both Ishii and the engineer, Hamauzu, explain that the structure was calculated and say that it needed to be that deep. As it is a public building the roof beam design was governed by the limited deflections of the timber members. After doing the calculations for the structure, it became apparent that there was an important requirement that governed the depth of the beams, one that went beyond the intention of the architect to have an oppressive and heavy structure in the space. It is clear that the architect and engineer were able to work very

▲ **7.53** The shop and café building – elevation. (Photo: Kazuhiro Ishii.)

▲ **7.54** The roof structure of the shop and café building is a type of RF structure – internal view. (Photo: Kazuhiro Ishii.)

closely from the early stages of the design. This teamwork of architect and engineer has resulted in a very beautifully crafted piece of architecture, where technical and aesthetic considerations are in full harmony.

The other two buildings on the site, although different, are equally successful. The shop and café are housed in an elongated and curved-in plan building whose roof truss uses RF principles. It is a truss where interlocking beams that are shorter than the span are used which, although different to an RF structure, has some resemblance to it. In a way, it is similar to the temporary bridges that Leonardo da Vinci designed (see Chapter 2).

▲ 7.55 Physical model of the café/shop roof structure used in the design process.

▲ 7.56 Façade detail.

The last building on the site, the restaurant, was erected in 2004. Although very different to the three buildings described so far, it is interesting how Ishii has been able to work with the same theme of grillage structures and develop it a stage further. The restaurant is housed in two volumes, each covered with a membrane structure. The load-bearing part of both buildings is a very unusual combination of rough, massive section, round timbers interlocked in a grillage structure with steel-pinned connections. The contrast of the rough timber and the smooth steel pins, combined with the lightness of the membrane, creates a

▲ **7.57** Restaurant building – external view. (Photo: Kazuhiro Ishii.)

▲ **7.58** Internal view of the restaurant. (Photo: Kazuhiro Ishii.)

▲ **7.59** The pinned timber structure folds and is locked into its final position. (Photo: Kazuhiro Ishii.)

magnificent space. And although the restaurant structure does not work like an RF, it takes the idea of the RF to another level of development, one that complements the architect's vision.

It is interesting that at the time of construction of the Seiwa Burnaku Puppet Theatre (it was built in 1994), it was necessary to make physical models of parts of the structure at 1:3 scale in order to convince the authorities and to make sure that all the complex joints would fit together. The building complex is like a huge three-dimensional jigsaw. All joints are carpentry joints and everything slots and fits together. The joints are all based on traditional Japanese joints that have been developed for this purpose. All the RF structures used in these buildings, with the exception of the auditorium 'chopstick grillage', are constructed without the use of any metal connectors. No nails or screws were used to put these great puzzles together. All materials including timber as well as construction workers were local, which helped the economy of Seiwa.

When I ask Ishii about the complex notching of the beams and whether any mistakes were made in the cutting, he simply replies:

> *'If they [the construction workers] had made mistakes, they hid them from me. I never heard about them. I know it was not easy to build the Seiwa Burnaku Theatre, but what is easy in life?'*

In my view The Seiwa Burnaku Puppet Theatre is one of the most remarkable applications of RF architecture. For me it is a building complex that synthesizes architecture and engineering in the most beautiful way. It is a design that is about unity of the old and the new, about dialogue, and about achieving form through the exploration of how to use materials and structures to tell a story, a story of architecture. After visiting this building, still beautiful though it was built in the early 1990s, I look at buildings in a different way. I expect more from them. I recommend the experience of visiting Seiwa and the beautiful RF structures by Kazuhiro Ishii to everyone.

8

Torikabuto – the Life Science Laboratory designed by Yoichi Kan

The reciprocal frame as an ecological structure

On Saturday 25 November 2006, I met Yoichi Kan at the railway station in Nagasaki, situated on Kyushu, the most westerly of Japan's main islands. He was accompanied by Mrs Keiko Miyahara, the wife of his close friend and colleague Mr Miyahara, who is a Professor of Architecture at Nagasaki Institute of Applied Science. Mrs Miyahara has a degree in English and her role was to aid in the communication between Yoichi Kan and myself. Kan speaks English, but he felt that it would still help to have Mrs Miyahara with us. We had agreed the schedule of my visit in advance and as planned we set off in a four-wheel drive on a journey to visit Kan's reciprocal frame (RF) design. It is a 50-minute drive up in the mountains near Omura, to the north-east of Nagasaki. As we climb higher and higher up the mountain, the road becomes more and more narrow until it is just a tiny, single-lane road. The surroundings are breathtaking. We drive through beautiful cedar forests and green fields. We can see the peaks of the Tara mountain range. It is a clear, early autumn day and the changing colours of Nature make the surroundings even more beautiful. We talk in the car about the book, my job at Sheffield University, my family, my trip and impressions of Japan so far. Straight away I feel at ease with these kind people, so I tell them how I am really impressed by Japan, that I find everything different to the Western world but because I feel people are so friendly and keen to help it is very difficult to get lost. I think they are relieved that I like Japanese food and seem amused that I find it a bit difficult to handle noodle soup with chopsticks.

During our journey I find out that Kan, who was trained as a structural engineer at Nagasaki University, comes from the island of Shikoku. He grew up in the countryside in a family of carpenters. Three generations of his family, his great-grandfather, grandfather and father, had been carpenters. He was always close to Nature and feels strongly connected to it. He is the Managing Director (the Japanese title is President) of Pal Corporation Group, a building design consultancy that employs about 65 people, including three doctors of engineering, 30 civil and structural engineers, 12 qualified architects and service engineers. Pal Corporation Group is a very successful and respected organization with a turnover of 3.4 million pounds (800 million yen). Although their work is mainly in Japan, in recent years they have been expanding overseas and have been involved in projects in many countries in Asia. Pal Corporation Group is involved in the structural design of many kinds of buildings, civil structures and power plant facilities based on the structural design codes of many countries in the world. Their projects also include structural analysis, establishing the strength of materials or mechanical systems as well as dealing with problems of vibration or fatigue in structures. Furthermore, they are involved in research and software development for technical calculations, as well as CAD systems.

The journey goes very quickly, and all of a sudden Kan stops the car: together with Mrs Miyahara I am invited to follow him. We are at the Life Sciences Laboratory, Torikabuto. I hear that 'Torikabuto' means cock's comb. The laboratory was named after the mountain at the back, Mount Torikabuto, which has three peaks that together resemble a cock's comb. On the other hand, the word 'Torikabuto' is widely known to the people in Japan as the name of a very poisonous plant. This plant is also called 'Torikabuto' because the shape of the flower looks like a cock's comb.

As we get out of the car I notice the reciprocal frame building that until then I had only seen in photos. It looks even more stunning than I could have imagined. However, to my surprise it is not the only building on the site – it is part of a whole complex of ecological structures that are set in landscaped gardens planted with healing herbs; there is a vegetable garden, a place for free range chickens, a Buckminster Fuller geodesic dome and an elongated building with photovoltaics. In the distance I notice a small wind turbine on a pyramid-like building that houses the toilet block. Another building nearby is the children's accommodation block.

At this point, Yoichi Kan invites us to enter the reciprocal frame building, the New Farmhouse as he calls it. We enter the building, which has a roof light at the top. The light comes both from the side windows and

▲ **8.1** Site plan drawing. (Drawing: Yoichi Kan.)

▲ **8.2** The new farmhouse building and the Fuller dome – hand sketch. (Drawing: Yoichi Kan.)

▲ **8.3** The designer of the New Farmhouse building – engineer Kan.

▲ **8.4** The RF building in its surroundings.

▲ **8.5** The entrance hall of the New Farmhouse building. (Photo: Yoichi Kan.)

▲ **8.6** The RF roof structure is exposed internally. (Photo: Yoichi Kan.)

the roof. The whole space is flooded in light. The building is square in plan and, following the Japanese traditional farmhouse design, it has four rooms divided by sliding partitions but no corridors. The spaces flow into one another and are formed by closing or opening the sliding partitions. We go one step up and enter a tatami room with a small table in the middle and few cushions on the floor. In the traditional way, we sit on the cushions on the tatami floor and have lovely Japanese cakes and green tea. The reciprocal frame roof is visible from all the spaces because there are no ceilings enclosing them. It is the reciprocal frame roof with the external walls that creates the enclosure. The greater than usual height for accommodation rooms and the lack of furniture makes the spaces feel larger than they really are. The spaces have a warm feeling because of the light that comes in and because of the use of natural materials.

The beautifully detailed cedar wood used for the structure of the building adds to the feeling of warmth.

The span of the New Farmhouse building is 8 metres. One of the first things I notice is that the eight RF beams that form the roof structure are quite small in section. Kan tells me that they are 15 cm wide and 30 cm deep. When I ask how this works, Kan explains that he calculated the beams so that they would take their own weight and the weight of the roof, including wind and snow loads. I ask about the notch between the beams, at which point Kan says, 'What notch? There isn't one!' This is unusual because most RF structures are formed by beams that on the outer end are supported by an external column or load-bearing wall. At the inner end the structure becomes self-supporting and stable by creating a closed circuit of beams that mutually support each other. In most cases the RF roof is created by notching the upper beam, which when placed on top of the lower beam locks into position and creates a stable roof structure (as presented in Chapter 7). This has some advantages, such as the possibility of pre-cutting the timber joints in a workshop and expressing this type of joint and making it part of the overall architectural expression.

▲ 8.7 The interior is flooded with light.

▲ 8.8 When the sliding windows are opened the landscape extends into the RF building.

However, the notched beam approach also has some disadvantages. By notching the beam at the point of highest shear stress (each beam contributes with its own weight in a point load applied to the supporting beam), the beam is weakened at the least desirable place, because of which greater beam sections are required to achieve the necessary load-bearing capacity. Obviously, this makes the structure less efficient, and although structural efficiency is not always (and should not be) the most important factor in deciding on the type of structure to be used for a particular building, one must agree that it is an important one to consider. Perhaps a more important implication is the overall architectural expression achieved when notched beams are used or not. In the first case the relatively larger sections needed for the RF beams will contribute to a heavier-looking structure. This, as shown in some of the other case studies, may be fully appropriate and justified for some RF buildings and may be part of the whole aesthetic expression and narrative of the particular building. In the same way, the lightness of the structure of the New Farmhouse is part of the architectural expression of this building.

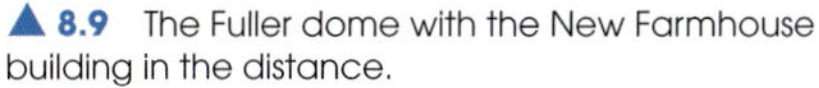

▲ **8.9** The Fuller dome with the New Farmhouse building in the distance.

▲ **8.10** The Fuller dome – close-up.

Another implication of using notched RF beams is the complexity of the joint and the need for very high precision, computer-aided design (CAD) and excellent carpentry skills. Because of the nature of the sloping RF beams, the notch has a quite a complex three-dimensional geometry which, if not cut to high precision, may lead to wasting of a slightly imperfectly pre-cut beam. Despite the computer calculations available to support this demanding carpentry task, there is no margin for error, and the risk of failure remains an issue to bear in mind.

But let us get back to Torikabuto. As we sit on the cushions on the floor of the New Farmhouse building and enjoy our green tea, Yoichi Kan starts telling me the story of how this amazing building and the Life Sciences Laboratory came to life. With the help of Mrs Miyahara, I hear the story in all its fine detail, the story of creating Torikabuto.

About 20 years ago, Kan was ill (suffering from gallstones) and had to spend 2 months in hospital. He had plenty of time to reflect on his life, read and think about the future. At this time he became very interested in ecology and ecological structures. The work of Frei Otto, among others, was an inspiration. Kan started thinking about more sustainable ways of living, how to reduce the waste we humans produce by reuse and recycling, how to utilize renewable energy and how to help future generations. He felt that he himself needed to have strength and energy to fulfil his role as a structural engineer. However, he knew that all people also needed good health and energy, and to get closer to Nature: living with Nature, he felt, was the only way of achieving this. This was when the 'seed of Torikabuto' was planted.

Yoichi Kan had an idea of creating a complex where issues that are the main concern of human society and the future lives of people should be explored. In his view, the main issues to explore were grouped broadly around five themes: natural and bio-structures; renewable energy; ecological design; human health and healing herbs; and the history and culture of the local community. He named the complex the Life Sciences Laboratory, and he envisaged it as a place where all these issues, which are of vital importance to humanity, would be explored. Also, he felt that he could help society if he could make future generations aware of all these issues and help them live in a more sustainable way, closer to Nature and with a healthier lifestyle. This is why he created an education centre for children as part of Torikabuto. It is interesting that Kan was the sole designer for the complex. He was not only the visionary but also the architect, structural engineer and planner; he did the landscape design and also acted as the project manager.

Soon after his recovery, Kan bought a piece of land at the foot of Mount Torikabuto. As he explains:

> *'Although I was seriously interested in exploring more sustainable ways of living, it all started as a purely business venture. An American company*

based in Japan had imported a geodesic dome structure for Tokyo and as the structure became very popular a Japanese developer based on the Amakusa Island in Kumamoto Prefecture wanted to start building them. The developer approached me and asked if my company, Pal Corporation, would be able to help with the structural and detail design. Here in Japan, normally the local authorities would approve the design of any structure, but in this case because it was a special structure we had to send our design to the central government body in Tokyo for approval. It took about a year for the government to grant approval and during this time I decided that it would be good to construct one of these structures on the mountain site I had bought. It was a novel structure and I wanted to explore novel construction methods. That is how the first building was erected. The structure of the Fuller dome is built of a timber frame, the inside is timber clad and the outside layer is created with plywood panels covered with a layer of FRP [fibre-reinforced plastic] for waterproofing.'

It is interesting that the Fuller dome (Figures 8.9 and 8.10) has a very traditional earth floor, one that would have been used probably 1000 years ago. The contrast between the perfectly laid timber-clad walls and the traditional earth floor are in complete harmony with the geodesic form of the Fuller structure. At the time I could not understand why the combination works so well. Yet by thinking about Nature and the very essence of this whole complex that Yoichi Kan has created it dawned on me: a Fuller dome is a natural structure and as such it reflects the laws of Nature where the perfect orders of geometry are combined with the randomness of chance and an ordered chaotic (looking) universe. Thus, the perfect geometry of the geodesic structure and the roughness of the stabilized earth floor work well together. They both are part of a bio-structure and part of Nature.

So, that is how it all started, and from then on the site grew as a laboratory of ecological research and design.

Kan continues his fascinating story:

'Very soon after the Fuller dome was completed, I started working on the design of the New Farmhouse, the RF residence. You must remember that by now I had a clear vision of what I wanted Torikabuto to become. I had a vision of the Life Sciences Laboratory and a research and education centre where we can both explore new and more sustainable ways of living, but also educate people about the great opportunities of this new approach. At this time [1993], when I designed the New Farmhouse I had done the site planning and we had planted the herbs, vegetable and fruit garden, brought in chickens that live on the site and had built some of the small buildings. I felt very fortunate that a small river forms one of the borders of the site, which provides water for the plants.'

The New Farmhouse was designed and built in 1993, and by talking to Kan I found out that at the time of its creation he knew of no other similar structures. His inspiration for it, and especially for the RF structure, came from three sources: he was inspired by the way baseball bats are sometimes assembled, by the traditional way of arranging agricultural tools and by Japanese forms of origami. It is interesting that RF structures have been designed by different people and although all of these structures share a very similar concept, the sources of inspiration for their creation have been derived from different ideas, phenomena, forms and objects. As seen in this and the other case studies, the RF structure has a multitude of meanings for the buildings it creates: it helps create different narratives and contributes in different ways to their architecture, as if it has the ability to represent itself in many faces and forms, a new one for each occasion and form of architecture it contributes to.

Through sketching, Kan decided on an eight-beam reciprocal frame roof for the New Farmhouse building. The architecture of the building was very much structure-led, because the RF structure was created first.

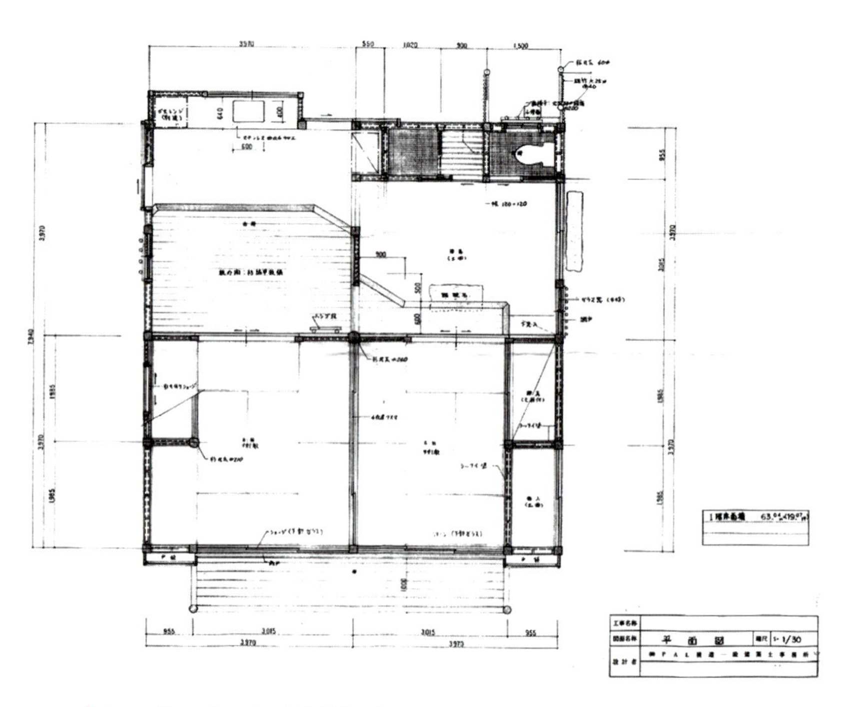

▲ **8.11** Plan. (Drawing: Yoichi Kan.)

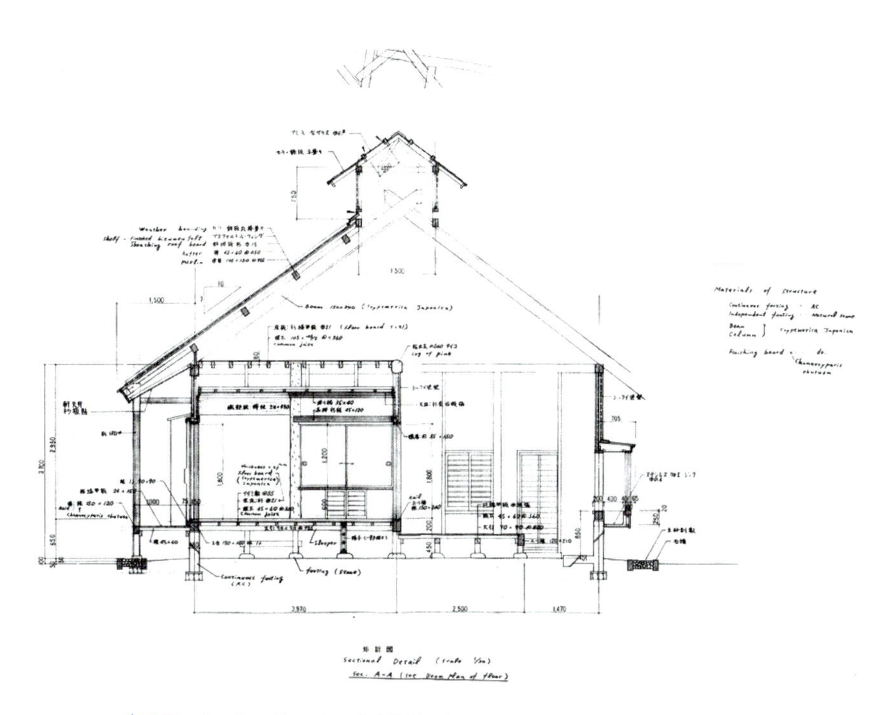

▲ **8.12** Section. (Drawing: Yoichi Kan.)

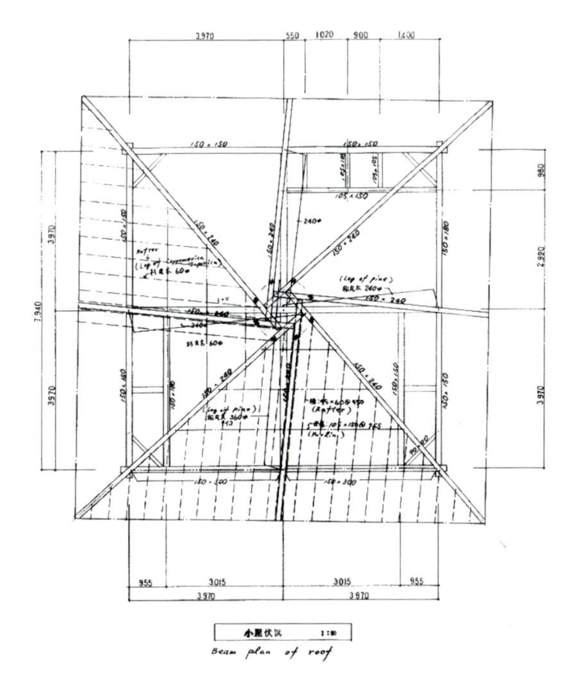

▲ **8.13** Roof plan. (Drawing: Yoichi Kan.)

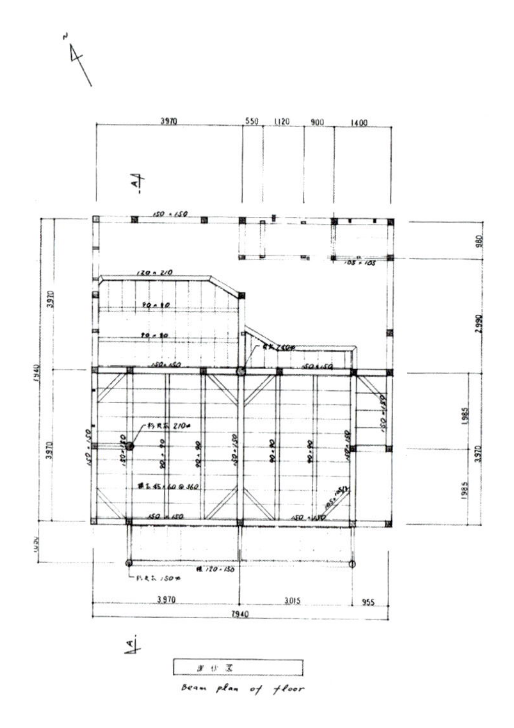

▲ 8.14 Floor plan. (Drawing: Yoichi Kan.)

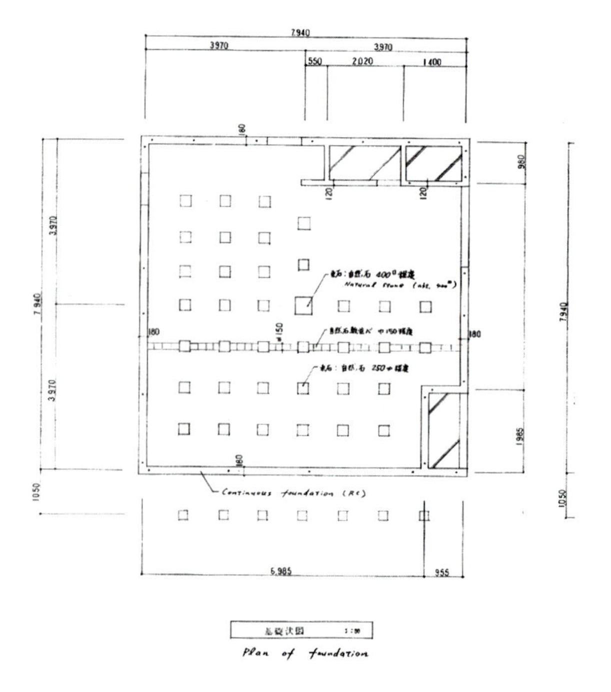

▲ 8.15 Foundation plan. (Drawing: Yoichi Kan.)

To a great degree the RF structure determines the architecture of the New Farmhouse building. As Kan says:

> *'The RF is an ecological structure. I see it as part of Nature. It is like the biological structures that Frei Otto writes about.'*

Its beauty is that it compares to the structures we find in Nature. In its concept it is as refined as the natural structures that we are surrounded with: beehives, trees, the stems of plants, the nerve structure of leaves, sea shells and so on.

As one of the themes of Torikabuto is local history and culture, Kan wanted to recreate the traditional farmhouse but in a new way. As he explains:

> *'Traditionally, farmhouses in Japan are square buildings in plan and quite dark. On the one hand, I wanted to create a traditional farmhouse but at the same time I wanted to create a building with spaces that have a lot of natural light. That is how I started thinking about a roof structure that could accommodate a roof window. On the other hand, as I wanted visitors to get to know the local history and culture, I chose to recreate the old traditional plan form of the farmhouse design. That is how the RF structure came about. You could say that it was created out of necessity, out of my strong belief in ecological structures and through my wish to explore and create new designs.'*

The RF roof is the distinct feature of the New Farmhouse building which makes the building special. It not only brings light from above to the four spaces, but also creates a feeling of lightness, achieved to a great extent by the floating beams that seem to touch very lightly towards their top end. The small timber sections that Kan has used in his design clearly contribute to this feeling of lightness. The daring structural design with the minimal RF beams and the overall quality of the detailing makes this building what it is, a beautiful example of RF architecture. It is a real synthesis of structure and architecture, achieving beauty through the clear understanding of structural principles.

Kan continues his story:

> *'After deciding on an eight-beam RF structure for the roof and a square plan for the building, I struggled to make them work together. It would have been considerably easier to have an octagonal plan form, but I felt that that would have been wrong considering that there are no octagonal farmhouses in Japan! The geometry was quite complex, the most difficult being the positioning of the RF beams in space. We used computer calculations and we also had to construct 1:5 physical models which we had to scale up*

to find the exact position of the beams in space. The building is a timber frame structure built in local cedar wood. It has nine timber posts and vertical bracing in the three vertical planes of the external walls. All the timber joints are done by local carpenters and are based on traditional joints. They are modified to suit this design, but traditional Japanese carpentry was a starting point in the connection design. None of them use any metal connectors.'

▲ **8.16** Traditional Japanese timber joints are used throughout the building. (Photo: Yoichi Kan.)

That is another thing that fascinates me. The beam–column joints look so complicated and the only way the posts and the beams could fit perfectly is to have all the timbers pre-cut to very high precision. I think Japanese carpentry skills are unrivalled in other countries.

I ask Kan about horizontal and dynamic loads, such as wind uplift and earthquakes. He explains that everything has been taken into account. The earthquake resistance is achieved by using timber joints that are able to move with the earthquake motion and dissipate the dynamic earthquake energy in that way. The wind uplift is also accounted for in the design and sizing of the RF timber beams. When I confront him by asking him whether the roof beams are not too lightweight, he admits that they did put a concealed bolt through the RF members as a double security against wind uplift. As everything was new and experimental he did not want to take risks. But he says that the roof would have been fine even without the bolt.

The internal spaces and the detailing are beautiful. They are done with meticulous precision. The sliding panels form the spaces both internally and externally.

▲ **8.17** Construction of the RF is done using a temporary prop.

▲ **8.18** The RF beams are positioned on the temporary prop. (Photo: Yoichi Kan.)

▲ **8.19** More beams are positioned around until a full circle is formed. (Photo: Yoichi Kan.)

▲ **8.20** The RF structure becomes stable when all the beams are installed. (Photo: Yoichi Kan.)

▲ **8.21** The RF beams form a full circle. (Photo: Yoichi Kan.)

▲ **8.22** View through the roof light. (Photo: Yoichi Kan.)

▲ **8.23** The New Farmhouse is a timber frame building. (Photo: Yoichi Kan.)

▲ **8.24** The roof is enclosed with metal sheet cladding. (Photo: Yoichi Kan.)

After designing this novel structure, Kan also had to devise a way of constructing it. The innovative design required an innovative means of construction (Figures 8.17–8.24).

The New Farmhouse was built out of local materials and using local construction workers. It was a challenge to construct the RF roof structure. To achieve that, he devised, together with his team of carpenters, an octahedral template made of plywood which allowed the top end of the RF beams to be easily placed in the correct position. As soon as the RF inner beam circle was complete, the template which acted as a support was removed. The RF beams had become self-supporting at this point.

The New Farmhouse is a timber frame building on concrete strip foundations. The glazed external walls slide in a similar way to Japanese traditional buildings. The glazing can be covered by paper sliding panels to prevent heat loss or partly covered to allow for light to come into the spaces whilst still keeping the heat inside. The external walls are finished with traditional lime plaster. The inner partitions also slide in the same way as in traditional Japanese buildings. This allows for a flexible use of the spaces. The building's finishes are beautifully done: it is a very special building to be in.

▲ **8.25** Outdoor bath.

▲ **8.26** Children's accommodation block.

▲ **8.27** Toilet pyramid building.

▲ **8.28** Every year, children from all over the country visit Torikabuto. (Photo: Yoichi Kan.)

Torikabuto is a complex of ecological buildings and natural structures set in the beautiful setting of the mountains and in the landscaped herb, vegetable and fruit gardens. Since the early 1990s, when the Fuller dome and the New Farmhouse building were constructed, Kan has experimented with using new technologies such as wind and solar power, as well as novel construction methods. The children's block, finished eight years ago, is for example created by converting three disused prefabricated containers which are connected externally with canopies created by using plastic sheets that are usually used for agricultural greenhouses. The toilet block is in the shape of a pyramid, with lighting provided by a small wind turbine placed at the top of the building. There are many more interesting structures and corners of the site to explore, such as the solar power plant facility, the charcoal burning furnace, the water-purifying system, some ponds for aquatic life, the mushroom growing logs, the 100 people outdoor rice cooker, the wooden platform in the river, the outdoor baths made of local stone.

Perhaps it is not only the RF New Farmhouse and its innovative design that I initially came to visit and learn about that make this place special.

It is the whole context and idea behind the creation of Torikabuto that are special. In the last 20 years Yoichi Kan has shared his knowledge about life sciences by running a non-profit summer school for children here. Every year, schoolchildren 7–14 years of age from all over Japan and from abroad come and stay at Torikabuto for a few weeks. At the Life Sciences Laboratory they learn about Nature and how to be part of it, about Japanese culture and traditions, about growing herbs, fruit and vegetables, renewable power, reusing and recycling, and about natural structures. Most importantly, they learn about the 'seed' that Kan has planted by creating Torikabuto, a seed that can grow and spread and maybe become a way of life for future generations. It can help them live in a more sustainable way and as part of Nature.

The Stonemason Museum by Yasufumi Kijima

During my reciprocal frame study trip of Japan, I was fortunate to visit the Toyoson Stonemason Museum by Yasufumi Kijima, designed in 1993. The roof structures of the two big, circular in plan, main volumes of the Museum are formed by using multiple reciprocal frame structures. As soon as one enters the exhibition hall of the building, these complex timber structures draw the attention of the visitor with the intricate way they hold up the roof.

▲ **9.1** The complex round-wood structure. (Photo: Keikaku-Inc.)

Unfortunately, I could not talk to the designer himself about how his design ideas for this project were developed, as Kijima sadly passed away in 1994. However, I was very pleased that Hiroshi Sawazaki, a colleague of Kijima and the Managing Director (in Japan the title is President) of Keikaku-Inc., agreed to talk to me.

Keikaku-Inc. was established in January 1971 by architects Yasufumi Kijima and Takefumi Aida. It is an architectural practice based in Tokyo that, over its 35 years of existence, has been involved in a great number of projects, including several Expo pavilions representing Japan at world expositions, schools, religious buildings, museums, hotels, industrial facilities as well as housing projects. The practice has been involved in many urban and town planning projects as well. The main ethos of the practice that Kijima and Aida established, Keikaku-Inc., was to design buildings that are closely connected with and growing out of the environment. These underlying ideas have been nourished and continued by the practice since Kijima's death.

The practice continues working and designing in an environmentally responsible way, and the main materials proposed in their projects are natural materials such as timber and bamboo. They are very keen to explore new forms of structure and architecture where traditional materials (timber and bamboo) are developed to a new level and used in novel ways. Their tensegrity timber forms, arched timber structures, grid shells and space frames create an original architectural language in timber. It is an architecture that is, to a great degree, influenced by the innovative structural form and is in full harmony with it.

Keikaku-Inc. as a practice is also very interested in the participation and involvement of clients and users in the design process. In many of their projects they have used the input of clients and users to arrive at the final architectural form of their buildings. Another interest of the practice is the investigation of how to create healing spaces by use of nature and natural materials.

Hiroshi Sawazaki worked with Kijima at Keikaku-Inc. for quite a few years before Kijima's death in 1994. He tells me about the Stonemason Museum and about Kijima the architect:

> *'Kijima was very interested in the work of Buckminster Fuller. For some time he had been studying his work and at the time before designing the Stonemason Museum he was in the process of finishing the translation of a book about Fuller's work. The structural form of the Stonemason Museum was directly influenced by the work of Buckminster Fuller.'*

I interrupt there by asking: 'But Buckminster Fuller never designed a structure that is similar to Kijima's roof structure at the Toyoson Museum. What is the link to Fuller's work?' Sawazaki explains:

'It is true that there is no similar structure designed by Buckminster Fuller that has a direct resemblance to the Stonemason Museum. It is more the way of thinking, the interest in novel structural forms and their relationship with architecture that were the influences on Kijima. Kijima, like Fuller, used physical models to explore new concepts. For the Stonemason Museum he made many physical models of the roof structure, some of which are still exhibited in the Museum. With the models he explored the relationship of the exposed roof structure and the space it enclosed.'

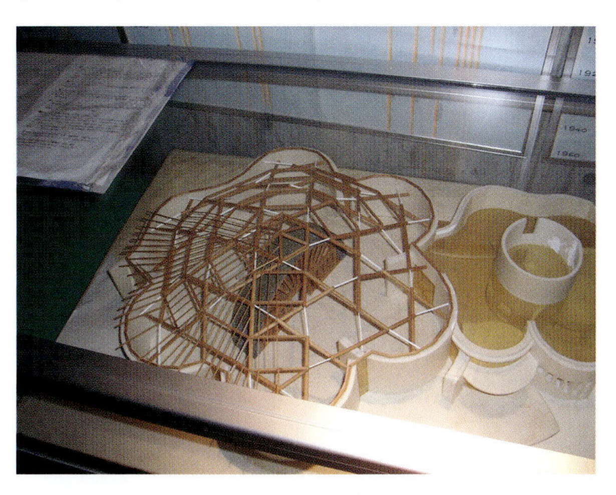

▲ **9.2** Kijima's physical model of the exhibition hall roof structure.

▲ **9.3** Kijima's physical model of the multi-purpose hall roof structure.

Hiroshi Sawazaki kindly gives me a monograph about the work of Yasufumi Kijima which covers Kijima's life in the period just after establishing Keikaku-Inc. 1972 to his death in 1994. With the help of a Japanese translator, I read about Kijima and understand him better as a person and as an architect. Unfortunately, despite his ground-breaking work in creating a new language of timber architecture through the integration of architecture with innovative structural forms, there is hardly anything written in English about his work. I have carefully read the few Japanese articles about his work (again with the help of a translator) and through this meeting with Sawazaki, as well as visiting the Stonemason Museum, I am hoping that I will understand Kijima's work sufficiently to describe his Stonemason Museum in the way he perceived it. I know that I am truly impressed by his building.

The Building

The Toh-yoh Village is a small community of approximately 3000 inhabitants set in a mountainous area of Yatsushiro County, Kumamoto Prefecture. This area is well known for its fine 'Shoh' (stone) material. The area has produced a group of fine stonemasons, some of whom are historically well known for their great works such as the Tsuh-jun Kyoh (or Tsuh-jun bridge), which is an aqueduct. In this area there are still some 22 stone-built bridges remaining: some are rather large and some are relatively small. The stonemasons from this part of Japan were well known for their masonry skills. They worked on Kumamoto's 'Tsuh-jun' bridge, Kagashima's 'Goishi' bridge and also the old 'Ni-juh' bridge of the Imperial Palace in Tokyo. There are also some manor houses with stone foundations which were built many years ago by the same stonemasons. Today, the village continues its traditional craft of stonemasonry and cares for the preservation of the heritage structures.

It is exactly here that the Toyoson Stonemason Museum was built. Kijima states in the monograph that the main aim of the Museum was to re-evaluate the traditional stonemasonry craft of the village and to promote stonemasonry using the Museum as an information centre. Thus, the Reference House of the Museum (Shiryoh-kan), which is the main exhibition hall, has some fine examples of Toh-yoh village's stonemasonry heritage.

The building is situated at the foot of high mountains on a sloped area which has been terraced. It is about 500 m away from the centre of the village. The Museum site is near the house that belonged to and was constructed by a famous family of stonemasons, the House of Hashimoto. Also nearby there are several old stone bridges that were constructed by the members of the same family. The permanent exhibition in the

▲ 9.4 The surroundings. (Photo: Keikaku-Inc.)

Museum shows small-scale physical models of local masonry bridges, as well as a 1:1 physical model showing the construction of a traditional stonemasonry arch. Kijima decided to build the Museum in stone, and several different forms and styles of stonemasonry can be seen in the wall construction of this building. It is an example of the fine art of the stonemasonry of this part of Japan.

The building with an area of approximately 800 m^2 consists of three distinct cylindrical volumes: the exhibition building and the multi-purpose hall, both of which have shallow metal-clad conical surface roofs, linked by the administration building which has a flat roof. The exhibition building houses all the changing exhibitions as well as the permanent exhibit, a 10-m-long replica of a stone arch bridge. The administration building consists of an entrance hall, an office and a café. The multi-purpose hall is used for meetings, lectures and some other functions.

When looking at the building externally, one cannot tell that the two main volumes, the exhibition building and the multi-purpose hall, have been formed by the use of multiple roof structures. The complex reciprocal frames are only visible to visitors when they enter the spaces.

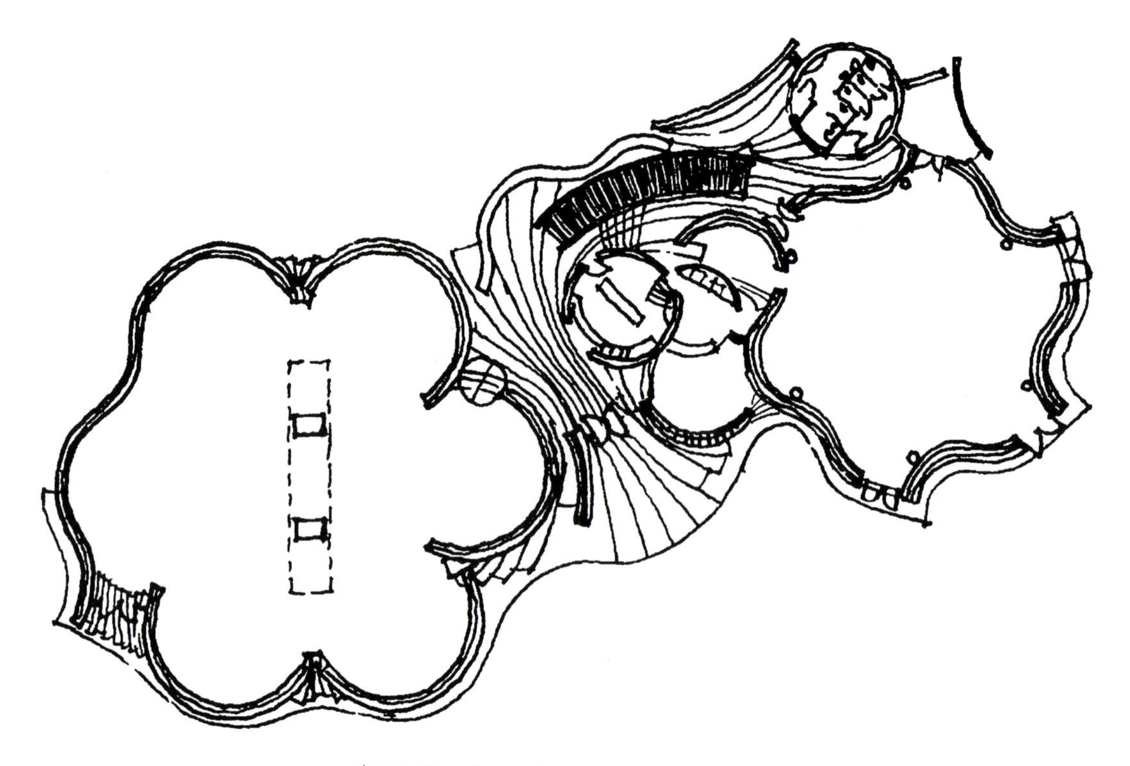

▲ **9.5** Plan: (from left) exhibition hall, admin block with café and multi-purpose hall. (Sketch by A. E. Piroozfar.)

The reciprocal frame (RF) structures that are used in this building are truly unique. On first view the exposed round-wood cypress poles look as if they have been arranged in a chaotic way: there are poles pointing in the most unexpected directions. Yet after just a few moments studying the roof, it is obvious that there is a clear hierarchy and that the pattern formed by the roof poles creates a very regular overlapping star-shaped arrangement. At the centre, the apex of the roof is a regular hexagonal RF unit which is supported by a combination of three-member single RF units combined with hexagonal RF units. Due to the geometrical characteristics of the multiple reciprocal frames, the 'in between' units are four-sided regular polygons in the shape of rhomboids that interlace between the triangles and the hexagons.

To direct visitors' attention to the exhibits, Kijima designed the natural lighting of this building very carefully. Between the semicircular niches formed by the external walls there is a narrow full-height window allowing a slit of light to enter the spaces. In addition, light comes in through the first circle of triangles formed by the three-member RFs.

▲ 9.6 The Stonemason Museum in context. The full height narrow windows bring in a slit of light.

It is interesting, though, that this is not direct light, but light let through the small clear storey windows positioned on the vertical walls of the elevated central hexagon of the roof.

The subdued light in the spaces and the unexpected roof structure of floating roof poles puzzles the visitor, who wonders how the roof structure over this whole open-plan space stands up. Not surprisingly, my eyes are drawn to the roof and I find myself following each roof beam, trying to work out which beam is supported by the other and which is supporting. It is an interwoven play of floating timber poles that, despite their chunkiness and size, appear lightweight. They are joined with metal connectors and metal bars which follow the star-like geometry. The roof structure forms a shallow dome form that, to a certain degree, is reminiscent of Leonardo da Vinci's early sketches (see Chapter 2).

▲ **9.7** Kijima's drawing of the roof structure configuration. (Photo: Keikaku-Inc.)

▲ **9.8** The Stonemason Museum in its surroundings. (Photo: Keikaku-Inc.)

▲ **9.9** View towards Toh-yoh.

▲ **9.10** Main entrance.

▲ **9.11** The roof under construction – external view. (Photo: Keikaku-Inc.)

▲ **9.12** The roof under construction – internal view. (Photo: Keikaku-Inc.)

In the exhibition hall there are several physical models constructed by Kijima himself. They are exhibited here to show visitors the architect's ideas. The model shows the RF roof structure clearly, with the timber and steel members that form it. In my conversation with Hiroshi Sawazaki, I find out that Kijima's idea was to have a timber-only roof structure. However, as it was impossible to make the roof structure stable by using timber alone. In discussion with his engineer Kijima he agreed to use steel bars which helped make the multiple RF structure stable. He felt that the only right thing to do in this case would be to be honest, showing plainly that two different materials have been used.

▲ 9.13 Interior view showing the permanent exhibit of a masonry bridge. (Photo: Keikaku-Inc.)

▲ 9.14 Roof structure – detail.

▲ 9.15 The craft of masonry is expressed in the building. (Photo: Keikaku-Inc.)

Thus, all the steel bars in the roof as well all the metal connectors are painted in a distinct red colour.

As the building is dedicated to the craft of masonry, the only right thing to do was to use stone for its construction. The building is thus on stone foundations. However, in order to fulfil the strict Japanese earthquake building design codes, the Stonemason Museum had to be constructed with a reinforced concrete frame as the load-bearing structure, using stone to create the external wall shell, which is not the primary load-bearing structure of the building.

All the stone for this building came from the local quarry, which was reopened for the construction of this building after many years of closure. Several local masons and three masons from China worked on the stone that was built into the walls of this building. Traditional construction methods were used wherever possible. Many centuries earlier, a great influence in masonry construction had came from China, so it was felt necessary to involve three Chinese masonry craftsmen in this project.

▲ **9.16** The intricate timber roof structure forms the space. (Photo: Keikaku-Inc.)

Kijima was an architect who strongly believed in the integration of structure and architecture. He had a great interest in how things are put together and how one can, as an architect, create structural forms that complement the overall design. During his working life he was both a practitioner at the practice (Keikaku-Inc.) he established in Tokyo and an academic: he was a professor at Kumamoto University, where he taught for over 20 years until his death. In addition, he was always at the cutting edge of architectural and engineering research. Just after his graduation at Waseda University in 1962, he went for 6 months to the Eduardo Torroja Research Institute in Madrid to research and study concrete shell structures and their application in architecture. Later in life he became a member of the International Association of Shell and Spatial Structures (IASS) and attended all the conferences, where he often talked about his cutting-edge design projects. He was a person with many interests. Kijima was a talented architect and, at the same time, someone who had a very technical mind. He could create the most amazing structural forms as part of his architecture.

Professor Mamoru Kawaguchi, the Vice President of the IASS, summarizes Kijima's life:

> *'For the 10 years until his death, Kijima attended all the conferences of the IASS and he tried his best to contribute to the development and the future of architecture. I now understand Kijima's attraction to the IASS was because he believed that it was a place in which all those who have their interests in and love for the development of the future of architecture could come and express themselves, regardless of their backgrounds, organizations or any academic empires.*
>
> *Thinking of this remarkable man Kijima, who loved and lived for his profession and its further development, I, too, humbly pray that the IASS will continue to nurture those who have the same passion as he had had towards our work and provide them a place where they can continue to discuss, debate and grow together in developing what we truly love: architecture.'*

10 The reciprocal frame as a spiritual structure – the work of Graham Brown

The First Meeting

I first heard of designer Graham Brown in the mid-1990s while studying the reciprocal frame as a Ph.D. student at Nottingham University. A few years prior to that Graham had come up with the reciprocal frame (RF) concept and had established a contact with the university, where research into the structural behaviour of the RF was started by Dr John Chilton. A couple of years later Graham had moved from Nottingham to Scotland with his family, and had set up his design and build company. He also started a timber workshop, where he became involved in timber fabrication. In Scotland, Graham was trying to establish himself as a reciprocal frame designer, which proved to be more difficult than expected: the problem was getting enough clients to commission him to design and build RF buildings.

For most building designs one would expect the structural concept to be developed to suit the design project and not be something that was pre-defined. Yet Graham, because he was fascinated with the RF concept which he had developed, was offering to build buildings with RF structures. This was a serious constraint. He not only needed clients, but he needed clients that wanted an RF building. Another limitation was that although Graham is a very talented designer and craftsman, he has no formal architectural education and has never been part of an architectural practice. All this made it difficult to attract clients. But Graham was building RFs and he was attracting clients.

Towards the end of my Ph.D. work in the summer of 1995, I went to visit his design office and wood workshop in Findhorn Bay in Scotland. I was amazed by his strong faith in the RF system and his enthusiasm.

▲ **10.1** Graham Brown constructing one of his RF buildings.

I found out that he was a qualified design engineer, acupuncturist, warm-cell installer and that in addition he had worked as a woodcraftsman, teacher and musician, but really the essence of his being was to create new things: he was a designer. Graham is an extraordinary person, one with a great spiritual depth, a person with great skills and many talents, and the one who gave the reciprocal frame, this extraordinary structure, its name.

At the time he showed me his completed RF buildings nearby. Also, together with Graham, I visited a local crafts fair where he had a stall: on the hour he was doing a small RF performance. He assembled a physical model of a house with an RF roof which he had previously made in his workshop. The unusual performance attracted the attention of the craft show visitors and always made a great impression when, after the completed assembly of the model house, Graham stood on the RF roof. Everyone was impressed that this small timber model could carry the weight of a man. He then talked about the structure, the quality of round spaces, the breathing walls he was proposing, and explained that he could design and build an exceptional house using the RF structure. With the strength of his faith in the system, his energy and persuasiveness, he managed to get people interested in commissioning him to design RF houses. Having in mind that a house is probably the biggest investment for most people over their lifetime, one must admire Graham's excellent skills in convincing potential clients. But despite Graham's faith in the RF, his design and timber construction skills, as well as persuasive abilities, life was not easy. Actually, in the early days of developing the RF it was a real struggle. Despite that, Graham has constructed over 30 RF buildings to date. He has had real difficulties, yet he has managed to fight to establish the RF. It is something he really believes in.

The Arrival of the RF

I met Graham again for the purpose of writing this book in October 2006. I visited him in his newly completed RF house in Findhorn Bay. I found that he had changed, as we all do, yet the strength of his passion for the RF had not altered. He was as enthusiastic as when I had first met him in 1995.

I asked Graham to tell me how he came up with the idea for the RF structure. He started his story by telling me that at the time he came up with the idea, he was working as an acupuncturist. He had trained for 3 years to become one and had been practising for 4 years. Yet what he was doing did not feel right. He continued:

> *'This was the first time that my spiritual life and my external life were aligned because I had been meditating by then for 10 years quite intensely. The feeling that I should be doing something else was very powerful, as if something was trying to knock on the door of my consciousness. It was for a year really, but by the end of it, it had got so strong that I thought I'd better stop. I need 3 months. I did not know why I thought that, why 3 months, but I thought I needed 3 months. However, I did not have money to have 3 months off so I took 1 month off. I did not have any money to go anywhere so I just sat there and wandered around the house (and got in Chris's[1] way).*
>
> *At some point I completely ran out of money and as it was coming up to Christmas I felt compelled to act on the basis of money, which I hated. I put up a notice in Sarah's[2] school in Nottingham that I would make toys, furniture and even a gazebo. I'd never made a gazebo before. From being a kid I used to make little shelters and loved it. I'd never built a building; I had renovated a house but to this day I do not know why I wrote that I would make a gazebo. It was interesting that Bob Pescar, a writer for Channel 4, whose children went to the same school, had seen my notice and asked me to build him a gazebo. I thought "Great!!" At a time when I am looking for a life-changing event, I suddenly get more work and that was not what I was looking for.*
>
> *I did not really want to do it and everywhere I went Bob was there asking me, "When are you going to come to see me about building this gazebo?" So one day I finally went to see him. I found out that in addition to the gazebo he wanted a studio building on top of his garage and also alterations to his*

[1] Chris is Graham's wife.

[2] Sarah is the eldest of Graham's three girls.

bedroom. All of this was thousands and thousands of pounds worth of work. And I had no money. I went home and remembered that he said one interesting thing: "I would like the gazebo to feel womb-like", which is a pretty unusual thing for a man to say.'

▲ **10.2** RF gazebo.

Graham went back home and started sketching. He drew a five-pointed interlocking star. As he drew the star he saw the possibility of taking one beam over the other and fixing it, just as the star does. He then got his daughter Sarah's pick-a-sticks and bluetak and tried to make a model. Very soon Graham found out that it did not work. He tried again, but this time he made a flat reciprocal frame. He thought: 'This is interesting!' and the following day he went to the workshop and made a physical model of the structure.

Graham continues:

'In the workshop I laid the first beam on the floor as I did with the pick-a-sticks but I made these 25 × 12 × 600 mm and I put the first one, then the next one and the next one and of course they build up and it was because I had done it with the pick-a-sticks I knew how to do it. So I held the last one in the air and picked up the first one and shaved under it.

▲ **10.3** Gazebo – internal view.

▲ **10.4** Gazebo – internal finishes.

> *And there it was: in three dimensions! So it wasn't that I designed it, it was as if it was waiting for me to discover it. And it was pulling me by the nose and saying "Would you please have a look at this!" That was very much how it was!'*

I interrupt by asking when was this happening. Graham tells me that it was back in 1987. He then continues his fascinating story:

> *'I had designed it and I knew that its geometry was very complex and yet on another level it was simple. So I needed to do it [build it] in order to work it out. So I started making a model of a building. This is on the same day it "actually arrived" because I was so hugely excited. I cannot describe that moment. It was just as if a small piece of God had landed on my lap. I looked at the completed RF structure and I pressed on it and it was strong. I stood on my newly constructed model and – it held me. This rush of joy in this Eureka moment flew through me. There was a voice that clearly said to me "This is a new structure; it is a new building structure; it is a new social structure; it is a new spiritual structure; it is a new financial structure – DO IT!" And that was it! That was the time when I made the structure in the workshop. And as I stood on it I got this complete rush of joy.'*

I ask Graham whether it was then that he formed his company 'Out of Nowhere'. He explains that that was later. When the RF 'arrived' he was still living in Nottingham. He was still designing his first gazebo for his client, Bob. Unfortunately, Bob moved away and he did not need the RF, so it was never built for him. Graham, however, became very interested in his creation. He realized that it was very complex geometrically because it is a three-dimensional structure. He spent 3 weeks building a very precise large-scale model and trying to understand how the structure worked. Graham remembers that time:

> *'I then realized that I had to do it really precisely and decided to do a really fine model. I soon realized that I needed to get points in space, so I started with the floor. I built this model very carefully and to do that I worked on it for about 3 weeks. I worked my way through it very patiently and it was very odd. I would go in and I would say to myself, "OK then, what am I supposed to do?" And it would just become clear. I just knew that I had to get my column details clear and with a circular building all the area of difficulty is in the column detailing, so I did that.'*

I interrupt by asking: 'Was the idea that it would always be circular?' Graham explains:

> *'It always was a circular impulse for me. It never came any other way. It has stayed that way. It is not that I have not drawn square things, but they do not have my energy. It was a bit of a surprise to me that Leonardo [da Vinci] had done some things and that there were some [RFs] in Japan. By then the RF had become something very personal for me and it has taken me a long time to understand what all that is about. While I was building the model it felt as if it unveiled itself to me. It was effortless. I still have it, it has been around the world with me and it was the only thing that survived the fire in my workshop.'*

The Patent Rights

It was in 1987 that Graham came up with his idea for an RF. He was so fascinated by it that he started talking to people about it. He thought it was a new invention, one that could offer a great deal. It was a time when he started collaborating with the University of Nottingham, where Dr John Chilton, Dr Ban Seng Choo and their students started investigating various aspects of it, such as the three-dimensional geometry, structural behaviour as well as the potential for using it as a retractable structure. I was looking at the architectural potential for the structure as well as investigating similar historical structures used throughout time.

The arrival of the RF was a life-changing event for Graham. He started to think about changing his profession and becoming an RF designer and builder. He shared his ideas about the RF with many people. On one hand he wanted to tell the world about it, on the other he was scared. The RF felt very personal, only his. He wanted to protect it. Graham explains:

> *'I then started talking to people about it. I showed it to John [Chilton] and others. It was then suggested to me to patent it. This felt wrong for me to do. But I realized that I was afraid. I was afraid of two things: it being misused and my work not being recognized. I gave in to this both external and internal pressure.'*

Graham contacted two patent agents. One of them, RGC Jenkinson and Company, based in Caxton Gate in London, invited Graham to their office in London. So Graham left Nottingham to meet Howard Millhench, one of the partners of RGC Jenkinson. He remembers:

> *'I went down to London from Nottingham but I had no idea where I was going. I came to the address and I saw an enormous building; it was a skyscraper really and RGC Jenkinson had all of it. Howard was on the top floor in the penthouse. I went into his private elevator to see him. He was a lovely Irish guy. We went into his office, where I took the RF sticks out, put them together and stood on the RF model. He looked at me and said: "Stay there! Put it down and wait. I am just going to get my partner." I put it up again for his partner to see and they were really fascinated. We were like kids playing with this thing. They said that it is eminently patentable, very easy to describe and that they could do it for me. I then asked "How much?", at which point they told me they could do it for £1000 easily. Probably the expression on my face made Howard offer a reduced price of £500. He explained that they very rarely worked with individual clients, so he was prepared to offer a 50% discount. That did not bring a smile to my face, actually I was looking deadly at them because I had no money. Howard said to me at that point: "So how much were you thinking of paying, nothing!?" Then I said: "Well, if you are offering…" They looked at me and said: "£250!" I replied, "OK, £250." So that started the process.'*

It is interesting that the investigations of the patent agents found no prior patents and no prior evidence of the idea, so Graham was granted patent rights for the UK, Canada and Australia. To extend the patent rights to the rest of Europe, Graham would have needed to spend another £6000–7000, at which point he decided not to continue. It was too expensive, but also for the whole time, deep down, Graham felt it was wrong to patent the structure.

I ask Graham about the benefit from getting patent rights: 'After you got the patent rights have people approached you to ask permission to use the RF, or for advice on how to design it and build it? Have you had any benefit of having patent rights at all? I've found that people have been building RFs and some of them mention you.' Graham explains:

'I had been used to people pinching my design ideas [it had happened before], so I felt fearful that I would lose the RF and I was protective of it, thus the patent. To advertise it I printed 1500 folders with technical information about the RF and in the space of 7 years they were distributed all over the world. I know that people who got hold of those brochures have built RFs.

Soon after the patent rights were granted, a friend of mine from Germany, Bertold, had been in touch with some people who had invested about £3 million in a timber machine that was very sophisticated and could do anything, but then they found that they did not have enough work. So they wanted to buy the rights for the RF from me. If I had sold them the copyright I would become bound to protect their rights. So if someone up the road built an RF I would be bound to litigate against them. I felt that if I said "yes" to that I would be saying "yes" to becoming a world policeman. It all felt wrong and I said "no". Shortly after that I realized that I had been overwhelmed by my fears and I had been pushed into actions that were wrong [to patent it]. I realized that I should not hold the RF and be protective of it. And in a way that is what I have done. It is out there. It is living a life of its own. So many people know about it. I have given birth to it and my duty now is to help people with it. I know a lot about it and I can give people information about what to do and what not to do.'

The Upward Struggle: from Gazebos and Whisky Barrels to Wimpey Homes

I ask Graham if, after distributing the RF information all over the world, he had many people getting in touch and if some of them had become his clients. Graham explains:

'I had thousands of enquiries that came through word of mouth. Hundreds of really nice people who had no money got in touch too. I spent loads of time talking to people about it. It made me understand the RF better and I become clearer about it, but it never made me any money. I had to do other things to support myself. Prior to 2000, when the Burial Park came along, I had built about 30 RF buildings: mainly small buildings, sanctuary buildings, two permaculture buildings in Bradford. At the time I thought this was the right thing to do. I just wanted to build RFs, but this nearly made me go bankrupt. I realized it was wrong. I felt that the financial system for the RF was not in place.'

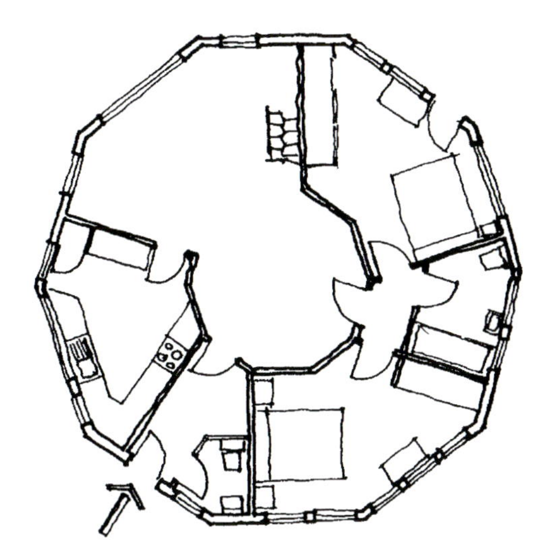

▲ **10.5** Ferryhill house – plan drawing. (Sketch by A. E. Piroozfar.)

▲ **10.6** Ferryhill house.

▲ **10.7** Ferryhill house – view towards the gallery.

The 30 RFs that Graham has built since 1988 have spans from 4.2 to 13 metres. All of them are in timber, solid timber for the shorter spans and glulam for the longer ones. With the exception of a few which have a circular plan, all the plan forms are seven- to 12-sided regular polygons. Among these are several houses, such as the two bedroom, 11-metre-diameter RF house at Ferryhill, near Forres, Morayshire, Scotland, and the 13-metre-diameter house, in Saorsa Ardlach, Nairn, Scotland. A recently completed project is Graham's private round house in Findhorn Bay. The other RF buildings are summer houses, gazebos and meditation retreats in private gardens. Also, in 1990, Graham was commissioned to design the RF structures for the roofs over two 6-metre-diameter whisky vats, both circular in plan, in order to provide living accommodation at the Findhorn Foundation, near Forres in Scotland. In 1995, the construction of three 8-metre RF modular pavilions designed as a Permaculture Centre in Bradford were constructed. His latest large-scale project is the Colney Wood burial park near Norwich.

▲ **10.8** The 13-metre-diameter house in Saorsa Ardlach.

I ask Graham why he always uses timber. He explains:

> *'I always work in timber because wood is a living material. I have built sanctuaries and when I've put the RF roof on and then put my hand onto them [the building] I've felt pulsing, low throbbing and aliveness. At first I could not believe it, but it is there! It has happened more than once. I work with materials that have "live" energy. Timber has it, stone has it too. I used concrete in my [round] house for the foundation and ground floors but they are supported by timber. It is a timber frame house with lime render.*

It has breathable walls and the concrete is an interface between the sun and the house. It is a balance that works. I was trained as a design engineer and in my early years I used a lot of steel, copper and other metals. I do not like steel. I go into a steel-frame building and I do not feel it is a place for humanity. It is a place for something else but certainly not a place for spirituality. I have no definitive reasons; I am just going by my feelings.'

During my last trip to Scotland, Graham took me to visit one of his earliest designs, a garden gazebo in Findhorn Bay (Figures 10.2–10.4). I had an opportunity to talk to the client, wood artist and craftsman, Richard Brockbank. I ask him why he decided to commission an RF gazebo design from Graham. Richard explains:

'We have four children and the house was getting too small, so we wanted to increase the volume of the building. The reason for choosing the RF was because Graham was at a stage when he was getting a lot of interest, but no one was saying "I want one now", so we wanted to help Graham build another one.'

Graham adds:

'Yes. It [the RF] originally started as a much bigger building. It was going to be a seven-sided 7-metre polygon and was going to be the guest space of the house. But then you built your extension and this became a much smaller building: it is 4.5 m span now.'

I ask if he has been happy using his RF gazebo? Richard replies:

'Yes. The only thing I regret is that we have not been able to use it to its full potential. I regret that. At the moment one of us occasionally sleeps there in the summer – it gets cold in winter. But we have not used it as a sanctuary and as a meeting space – we have not used it to its full potential. But it is beautiful. It is a lovely place to sleep in. I had someone staying there in September and she absolutely loved it. If we could have a slightly bigger one with a bit of a kitchen, toilet and a proper heating system, I would move out of the house and stay there.'

I look at the gazebo. It is one of the smallest RF buildings that Graham has designed, yet it is one of the most beautiful. The detailing of the wood shows Graham's great timber craftsmanship abilities. Also, the proportions are right: the shallow sloping RF roof and ratio of height to span of the gazebo work very well. It is not surprising that the client I talked to, Richard Brockbank, is very happy with the design. It is a truly successful RF design.

We get back into Graham's car and he takes us to the Findhorn Foundation. It is a typical October day in Scotland: it is raining. Yet it is very lively at Findhorn. The Foundation is a very interesting place. It was formed in 1962 by Peter and Eileen Caddy and Dorothy Maclean, who had followed disciplined spiritual paths for many years. They first came to north-east Scotland in 1957 to manage the run-down Cluny Hill Hotel in the town of Forres, which they did remarkably successfully. Eileen received guidance in her meditations from an inner divine source she called 'the still small voice within'. After several years, however, Peter and Eileen's employment was terminated, and with nowhere to go and little money, they moved with their three young sons and Dorothy to a caravan in the nearby seaside village of Findhorn. There they formed a community.

Since then, it has become a temporary and permanent home to people who are interested in alternative ways of living and who share common beliefs about sustainable living, including growing their own food and using less of the world's resources. In addition, Findhorn attracts people who have a very profound spirituality.

At present, the Findhorn Foundation is the educational and organizational cornerstone of the Findhorn Community, and its work is based on the values of planetary service, co-creation with nature and attunement to the divinity within all beings. The community members believe that humanity is engaged in an evolutionary expansion of consciousness, and seek to develop new ways of living infused with spiritual values.

Every year, people from all over the world come to learn about sustainable ways of living: about reusing and recycling; about bio, wind and solar energy generation; about food production; about various crafts such as wood crafting and pottery, as well as stone carving. People also come to deepen and develop their spirituality. Both architecturally and socially the place is a curious mix. It is an alternative community. On one level there are people who live in dilapidated caravans, cycle around on their scruffy bicycles and grow their own food. Yet there are others who have cars and live in the newly built experimental 'zero-energy' houses. Strangely, and despite the differences, it somehow seems to work. It may be because all the inhabitants and visitors share a common belief in sustainability and are people who also share a deep spirituality. It is here, at the Findhorn Foundation, that Graham was commissioned to build the whisky barrel RF roofs in 1990. I ask Graham to tell me more about them.

Graham explains:

> *'Roger Daudner heard that these whisky barrels were available at virtually no cost and it was an attempt to achieve a very-low-cost housing solution. It worked to a certain degree, but it became apparent that a great amount*

▲ **10.9** Whisky barrel RF house at the Findhorn Foundation.

▲ **10.10** Whisky barrel RF house – internal view.

▲ **10.11** Whisky barrel RF house – view towards the roof light.

of labour was needed. That was what the [Findhorn] Foundation always had available, but if you have a cost for labour, it works out to be an expensive thing.'

The whisky barrel RF roofs are clad with copper and, as with all Graham's buildings, express the RF both externally and internally. The copper cladding and the turbine-like stepped RF roof do not seem to go very happily together. It is a lightweight roof, yet it looks rather heavy. On the other hand, internally the roof forms a very beautiful space. It encloses an open-plan round house with a sleeping gallery, under which the kitchen and bathroom are positioned. The RF structure is expressed internally and when lying on the high level bed one can see a glimpse of the cosmos, very much as you can in Ishii's 'Spinning' house.

This project is a real milestone on Graham's RF journey. It is the first project where the structural behaviour, geometry and detailing were established by calculations. Before this project it had all been trial and error. Graham worked with structural engineer John Chilton, who did the structural and detail design for this project. John Chilton explains:

'My first RF building was the Findhorn recycled whisky barrel house. This was a building where we had to work out how the structure works, how to make and cut the notches and how to construct the roof. It was really exciting that when we pre-cut the RF beams and put them up, everything fitted together. Before this project it had been trial and error. This was the first project where the notch was designed.'

As we continue to the café at Findhorn, seeking a shelter from the rain that has become heavier, I ask Graham if he should have thought of patenting a 'flat pack' Graham Brown RF building instead of protecting the RF principle. He explains:

'I actually did it. I was commissioned to do a project for Wimpey Homes. They wanted a sales office and crèche. It was an eight-sided polygon with four windows and four solid external walls. I made the two small buildings for a relatively low cost, still managing to make a small profit. But it was hard: I had to talk to some hard-nosed businessmen. They paid me 80% when I delivered it on site, so I had to finance it all myself. But I did it and they were impressed by it. We put it up in a week and a half on site. And these [RF] houses were selling when they could not sell their own designs. But it felt wrong for me. For me the RF is an architectural mandala which is about the journey "home", whatever you want to call it: God, spiritual home or something else. It has energy of its own and it affects people's lives. In the time I have lived in my [new] RF house my life has completely been "undone", but I have been ready for it. I have observed other RF owners and many of them have gone through turmoil. Their lives have been changed completely.'

I ask: 'So you are saying that there is more than a flat pack to it?' Graham explains:

'Yes. It is its spirituality and its energy that are special. The RF creates a place of "breath" – without breath we would die, it is a generator of our exist-ence. For me it is as refined as that and when I was trying to make it into a commercial structure I was going against my own understanding of it. That was why it was not working. I was building RFs but deep inside I did not want it to be a commercial thing. I wanted the [building] experience but I did not want the RF to be put out of its context. It has taken me 15 years to understand that. After saying all this I feel my complete alignment. Nothing is standing between me and the RF apart from the fear of how am I going to make a living. I have a family and a responsibility to my family. But I've always felt that this is my life – the life of a nomad who is travelling and talking to people [about the RF].'

The RF as a Spiritual Structure – Colney Wood Burial Park

Throughout his RF journey, Graham always felt that the RF should be used as a spiritual structure. He has designed several sanctuaries using the RF. One was a temporary structure for the Earth sanctuary for the Eco-village conference at Findhorn. He talks about the experience:

'We put up the Earth sanctuary in four afternoons. The space is under-ground and only the RF roof is above ground. We used round wooden poles for the RF beams. It was Craig Gibson's inspiration [Craig lives in the whisky barrel RF in Findhorn]. We used larch poles and the connection is very sim-ple: there is only one bolt through them. We used a scallop notch to stop them [the beams] sliding on each other.'

The first afternoon in my workshop we prepared the beams [pre-cut them], the second afternoon we excavated the existing pit, the third afternoon we put the beams up. It was an amazing experience, with the women singing and blessing every beam by rocking it gently. When we needed to put in the last beam all of us went in and lifted the structure two inches and it all fitted perfectly. The roof is clad with timber planks that were nailed into the RF. A very thin membrane went over it and then the turf roof was put up. It was meant to be a temporary building that would be there only for the duration of the conference. But it stayed for 2 or 3 years until eventually the membrane gave in and it was taken down. But it has been replaced with another RF with a tent structure over it, so it is still there [at Findhorn].'

Graham's strong belief in the RF as spiritual structure, built in a sustain-able way, found an application in his latest large-scale project: the Colney Wood burial park. Colney Wood is near Norwich and offers a novel

▲ **10.12** Findhorn Earth sanctuary. Wooden poles were used for the RF roof.

concept for burials. In this complex there are three big RF buildings that are set in the beautifully landscaped woodlands: an office building, a gathering hall and a chapel. In addition, there are several small RFs, including a small shelter and a canopy at the entrance showing the setting of the park. Graham was the architect for all the RF buildings. John Chilton, structural engineer and professor at the Lincoln School of Architecture, did the structural design for the office building, whereas Peter Murray of Leonard Murray Associates from Nottingham designed the structure for the gathering hall and the chapel. It is one of the best RF designs that Graham has created. It is not surprising that it won the new building category in the South Norfolk Design Awards in 2004. The competition judges praised it as a 'highly sustainable design that harmonized with the woodland setting'.

Graham tells me the story of how he got the commission to design Colney Wood:

> *'At one time I put the small buildings in a publication* External Works, *and it is through that route that I have been commissioned to do the Woodland Burial Parks. The landscape architect John Dejardin saw the publication,*

▲ **10.13** Colney Wood Chapel.

▲ **10.14** Chapel – internal view.

▲ **10.15** Chapel - the building blends into the woodlands.

▲ **10.16** The coffin is carried out through the glass doors.

was interested but at the time [1995] nothing happened. In 2000, after my workshop had burnt down, I had some very difficult financial years. A man called Donald Body, who had heard about me from the landscape architect, got in touch and wanted a building. I thought: Donald Body *from a* burial park *– this is one of my mates having a bit of a laugh! So I ignored it, but then they gave me another call [him and his brother] and I realized it was real. It is interesting that the commission came through the publication,* External Works, *yet that was the only place I had ever advertised it [the RF]!'*

Graham continues:

'I really believed in this project because I always felt that a sanctuary was the right way to go. I went down to see them and they were interested. Donald and John asked many questions and wanted us to put together a financial plan. The rough estimates about costing somehow got cast into stone. After 6 months of talking and my free advice to them they invited me to participate in an open architectural competition. I was furious and scared – I was thinking they would get a whole bunch of architects and what am I going to do? But I believed so much in the project that I was going to do it. They wanted me to do it for free, but then we agreed that they would give me half the fee and, if I were to win the competition, the

▲ **10.17** Chapel roof detail.

▲ **10.18** Eaves detail – side view.

other half with a £2000 bonus. I said OK. I had 6 weeks to meditate. After 4 weeks I was still waiting for the big idea to come and nothing had happened. I had lots of ideas for organic forms but none of them worked. And then all of a sudden this idea came to me about this simple linear geometry and I could see it clearly as a journey to the burial. On that journey there were places to stop where you needed to do things to honour the last journey of the deceased; there were seven gates and the last stop was the grave. So the park was designed on that concept and the buildings were designed around it. There are several buildings. Everything works with a simple geometry which makes it all very refined. The gathering building forms

▲ **10.19** Eaves detail – front view.

▲ **10.20** Roof light.

▲ **10.21** Roof light – detail.

a courtyard of about 20 metres. There, people are saying "hello" to others that they have not seen for some time. You can see clearly the ceremonial building, which makes you aware that soon you will need to go across the courtyard on a journey so that you can say your farewell to the person who has died.

The gathering building was designed for 80 people and for bigger funerals it opens out into the courtyard, from which it is separated by a living hazel wall. The coffin is carried to the ceremonial building and put under the RF roof light. People stand with stoicism, grief, love, pain, dignity next to the coffin with their beloved in it.

The building has three glass walls to allow the woodland to come into the building and also to tell people where the body is going to. It is a very hard moment when we realize that this body that we have loved is now lifeless and has to go on its journey to the woodland park. The ceremonial building is a gateway and the roof light is a gateway to heaven. It gives us the chance to understand that the love for this person who has died here in this building has to be freed from the body. It is a terrible moment for any human but also a moment of great beauty if it can be embraced. And that is what this building is about: it informs you quietly about what is to occur.

The coffin is carried through the glass doors through the woodland and then to an elevated place before it is buried. And after this people go back to the gathering hall. The first funeral I attended there made me realize something I could not imagine. After the burial, when people went back to the gathering hall, their lives started again. I could have not imagined this. The people have the opportunity to deal with their grief by being alone, together in the buildings and in the woodland. The place helps them. After the first funeral, with tears in her eyes, the widow of the deceased man said to me, "It is weird to say, but we have had a great day." I knew that she had the opportunity to do what she needed to do to honour the life of her husband. I also knew that I had designed my best piece of work.'

In many ways this is true, especially about the chapel – or as Graham describes it – the sanctuary. Of all three buildings on the site and of all the RFs that Graham has designed to date, this is the only truly different one. The curved glued laminated timber beams and the slightly tilted roof light bring a special elegance to the building. The building works exactly how Graham envisaged it. It is a gateway to heaven and a place of both departure and beginning. It is not surprising that it has been awarded the main design prize (Figures 10.13–10.21).

The gathering hall also works well. It is a conventional polygonal RF building and its positioning is particularly successful. Together with the chapel and the courtyard, they help the journey exactly in the way that Graham envisaged.

▲ 10.22 Gathering building – front elevation.

▲ 10.23 Gathering building – elevation facing the internal courtyard.

▲ 10.24 Gathering building – internal view.

▲ **10.25** Internal view towards the roof light.

▲ **10.26** RF roof – close-up.

Although the office block is a reciprocal frame building and therefore complements the overall design idea by following the project's geometrical forms, it is the least successful building. Externally, in order to mark the entrance, it extends one of the RF members, which although it serves its function also makes the building look slightly odd. Internally, while the open-plan office works exceptionally well, the small subdivided offices and meeting rooms have unusual shapes. Once again, this shows that RFs work very well for open-plan functions or over symmetrical spaces. It is more difficult to create subdivided spaces within the polygonal geometry.

When one looks at Graham's RF buildings in Colney Wood it is obvious that his detailing of the eaves is done in a way to attract the eye to the

▲ **10.27** Office building – side elevation.

▲ **10.28** Office building – front elevation.

turbine-like RF roof. It is a specific aesthetic and one that brings different levels of appreciation for different people. I feel that because of this detailing the building gives the false impression from a distance that the RF beams are very deep and heavy, though actually they are not.

▲ **10.29** Office building – close-up.

▲ **10.30** Office building – internal view.

▲ **10.31** Office building – roof close-up.

▲ **10.32** The tranquil setting of Colney Wood.

▲ **10.33** Site plan, with the RF structures scattered in the wood.

▲ **10.34** A small shelter RF building.

▲ **10.35** The entrance map is in an inverted RF structure.

I visited the building in May 2007 and was touched by the tranquil setting and the landscape design, but most of all by the sanctuary RF building. When talking to the employees I find out that they are very happy to inhabit and work in Graham's buildings. They are convinced that it is mainly because of the design of the park that Colney Wood has been awarded the 'Cemetery of the Year' in the woodland cemeteries section for two years running. I find out that Graham's design for a new burial park is starting on a site in Epping in September, with the aim of having it completed by October 2008. In addition, the plan is to develop two more sites using Graham's designs in the next 10 years.

To me, the success of Colney Wood comes as no surprise. Graham has worked very hard to develop his RF designs to the level of the Colney Wood sanctuary building, which is a really beautiful piece of architecture.

In the future, Graham thinks that he should develop his work to help people in areas of conflict. He is hoping to develop community buildings such as sanctuaries and gathering spaces with the RF design and use his designs to aid peace building.

I hope that he will continue his journey of growth and development and will explore new RF building forms with different roof slopes, cladding and detailing.

Graham believes that the RF can bring peace into our hearts. After visiting Colney Wood, I do too. However, I also believe that the RF has so much more to offer architecturally.

11 BUILT EXAMPLES

In addition to the work of designers Ishii, Kan, Kijima and Brown presented in the reciprocal frame architecture case studies (Chapters 7–10), there are also other designers who have used reciprocal frames (RFs) in their building designs. Over the last 10 years, many of them have approached the author of this book seeking help on how to approach the design and construction of RFs. In addition to the designs known to the author, some of which have been built and some that have not, there must be others scattered all over the world that are to still to be discovered. Not using a common name for the reciprocal frame structure makes this an onerous task.

This chapter presents three domestic RF buildings. All of them are interesting architecturally and their designers share a vision of creating environmentally sustainable designs. They differ between each other in the way they have approached the application of the RF in their designs. Also, they are different from the Japanese examples in scale and function, as they are not public buildings.

The Roundhouse, designed by self-builder Tony Wrench, used only locally available sustainable materials, such as straw bales and wood, that were grown locally. It is a low-cost, self-built house which was completed on an extremely low budget of £3000. Not many contemporary houses can claim that.

The Deborah Gunn Residence in Virginia, designed by architect Fred Oesch, of Oesch Environmental Design, was a close collaboration between the owner, the architect and the builder which resulted in a design with strong green credentials. It is an autonomous house designed to be energy independent. As such, it has no connection to the electrical power grid.

The Spey Valley reciprocal frame house designed by young Australian architect Hugh Adamson whilst working in Scotland is located in a

National Park. Consequently, it was difficult to get planning consent for this unusual RF building design. However, the architect and client, working closely together, managed to get planning permission to complete the house. This project, considered in detail later in the chapter, also uses several RF design elements that are the first of their kind.

The Roundhouse

Background

The Roundhouse is located in the small community Brithdir Mawr in Pembrokeshire in south-west Wales. It was designed and constructed in 1997 by Tony Wrench and his partner Jane Faith, who have been living there since.

The main feature of the house is that it was designed and built to have a low impact on the environment. The design brief was to construct a dwelling that was sustainable, zero energy and in harmony with the surrounding woodlands. It was entirely a self-built project using mainly local materials and labour.

In Tony Wrench's opinion, it is very difficult to construct new buildings in the countryside in Britain because, as he writes in his book about the building, 'At the heart of our planning laws is the unspoken assumption that people and the countryside are bad for each other.' However, as he writes, 'it is as natural for us to build an appropriate shelter as it is for badgers' (www.thatroundhouse.info).

Tony Wrench did not seek planning permission, as he expected this would not be granted. Instead, the local authorities found out about the building in 1999, two years after it was completed. Since then, the threat of demolition has been hanging over it.

The building

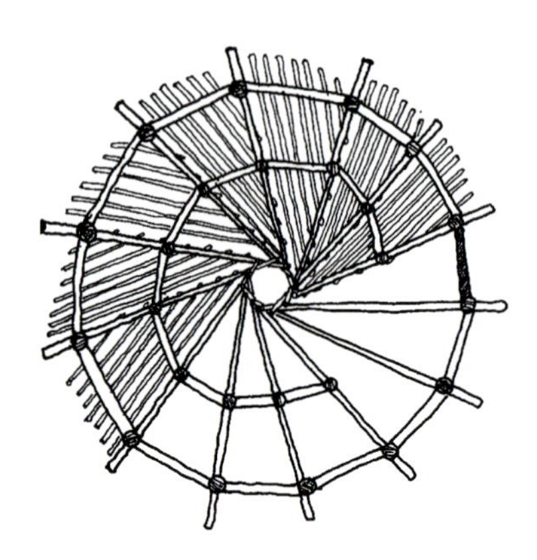

▲ 11.1 Roundhouse RF roof – plan. (Sketch by A. E. Piroozfar.)

The house, which is circular in plan and approximately 12 metres in diameter, is constructed from round wood, Douglas Fir logs, which were used for the columns and roof members. The timber poles, approximately 225 mm (9 in) in diameter, are positioned in the circular plan at approximately 2–3 m centres.

The roof is of reciprocal frame construction and consists of offset radial rafters spanning from the top of each column to the inner end, where they are supported by each other. Additional internal columns were inserted approximately halfway between the external columns and the inner end of the main rafters after the main frame was erected, in order to provide additional support for the main rafters. This helps carry the relatively heavy weight of the turf roof, which is distributed to each rafter. The supplementary vertical supports also provide restraint for the internal partitions that were put up later.

The depth of the main roof rafters is approximately 75 mm. The overall diameter of the house is approximately 12 m, whereas the diameter of the roof light is approximately 1.5 m. The roof light was enclosed with two large coach windscreens.

▲ **11.2** Roundhouse – internal view. (Photo: Tony Wrench.)

The roof is constructed from willow branches laid on the main rafters, and intermediate secondary rafters. Above this there is a canvas supporting straw bales, which are 300 mm deep, then a waterproof membrane (in this case rubber). The roof finish is a layer of turf on newspapers.

Horizontal eaves beams were laid on top of the columns, connected together with horizontal half-lap, pegged joints, to form a ring beam, which is capable of resisting the lateral thrust forces from the rafters.

▲ **11.3** Constructing the house – view showing the temporary support. (Photo: Tony Wrench.)

▲ **11.4** Constructing the roof. (Photo: Tony Wrench.)

The floor finish of the house is of packed earth construction. Concrete has not been used anywhere in the construction of the house.

The outer walls were constructed as infill panels between the columns, which are approximately 400 mm deep and made from a short length of round logs stacked on top of a damp-proof course. Cob and straw were used as mortar for the walls. Diagonal bracing members in timber were installed in some of the wall panels to the rear of the building.

The building is fitted with solar photovoltaic panels supplying the power. There is no mains connection. It has a grey water treatment system based on reed beds.

A woodstove connected to a hot water tank made out of a 270-litre brandy barrel (!) provides both fabric heating and hot domestic water. Sanitary facilities are provided outside the building.

Design process

The design process for the Roundhouse was completely informal and evolved with the design. The concept for the house was established from very early on, though detailed design decisions were left until late into the construction period.

The aim of the design was to provide a self-contained, low-energy dwelling with minimum carbon footprint and low impact on the immediate environment, with construction costs kept to an absolute minimum. It is clear that the finished building does not provide the quality of finish or comfort that is found in common houses in the UK, but, as Tony Wrench writes, 'the house has more in common with a shack in a shanty town in Buenos Aires than it does with a new Wimpey house in England'.

The construction of the building was largely self-built, relying to a high degree on friends. Tony Wrench is not a qualified architect or builder, but for this project he was the architect, engineer, client and builder. Tony has a strong interest in ecological issues. His brief to himself, written down at an early stage of the process, was simply: 'An autonomous house of wood, very warm, very dry, cheap to run. Made from pine logs from Erw Deg. Turf/bracken roof . . . it is built on a slope near woods.' He has certainly achieved that.

Construction of the reciprocal frame

The erection of the roof structure was carried out using a so-called Charlie stick, which is a temporary, central post used to support the roof rafters until all the main rafters are in place when the Charlie stick can be removed.

▲ **11.5** The finished skeleton. (Photo: Tony Wrench.)

All the columns were positioned in holes, hand dug in the ground.

The construction of the building took only 4 months. The total building costs were £3000, spent mainly on materials.

Deborah Gunn Residence, Virginia, USA

Background

The building is a private cottage built for its owner, Deborah Gunn. It was designed and constructed between 2005 and 2007. The project started as a close collaboration between the owner, architect and builder. The aim was to design and build a 'small, affordable, healthful, zero-energy home' (www.fredoesch.com).

Fred Oesch, of Oesch Environmental Design, Virginia, was the architect of the house and the builder was Bruce Guss, of Housewright, Virginia.

The building

The Deborah Gunn residence is located in a rural, woodland setting in Virginia, USA. The house is on two floors with a total floor area of approximately 120 square metres (1236 square feet). The roof of the building is of reciprocal frame construction. It is a timber frame house constructed from engineered glued laminated beams and locally sourced oak timber. The lower floor forms a concrete podium structure on which the timber frame of the upper floor is erected.

The house is octagonal in plan. The reciprocal frame that creates the roof enclosure of the building has eight main rafters arranged around a central roof light with a diameter of approximately 2.6 metres (8 feet). The overall clear span of the roof structure is 8.6 metres (26 feet).

For the roof, horizontal eaves beams were positioned on top of the columns, connected together with horizontal lapped joints forming a ring beam, which is capable of resisting the lateral thrust forces from the reciprocal frame main rafters. Secondary rafters were positioned over the main rafters to support the roof cladding which, in the case of the Deborah Gunn residence, is a vegetated living roof.

The outer walls are constructed from 300-mm (1-foot)-diameter poplar corner posts, with 600 mm × 1800 mm (2 × 6 feet) conventional wall framing, and straw bale infill.

The house was designed to be energy independent and so has no connection to the electrical power grid. Its power supply comes from 15 150-watt solar panels mounted on a nearby storage building. This system is supported by a back-up generator which starts automatically when the solar battery levels drop under a certain limit. Radiant heating and domestic hot water are produced by a high-efficiency liquid petroleum gas boiler supplied through an in-floor hydronic heating system. There is heating back-up provided by both a passive solar system and a wood-burning stove.

▲ **11.6** External view of the house. (Drawing: Fred Oesch.)

Construction of the reciprocal frame

The roof structure was assembled on the sub-floor platform and then lifted into place on top of the columns by crane. The roof assembly included the main rafters, the eaves ring beam, the secondary rafters and the timber frame for the central roof light.

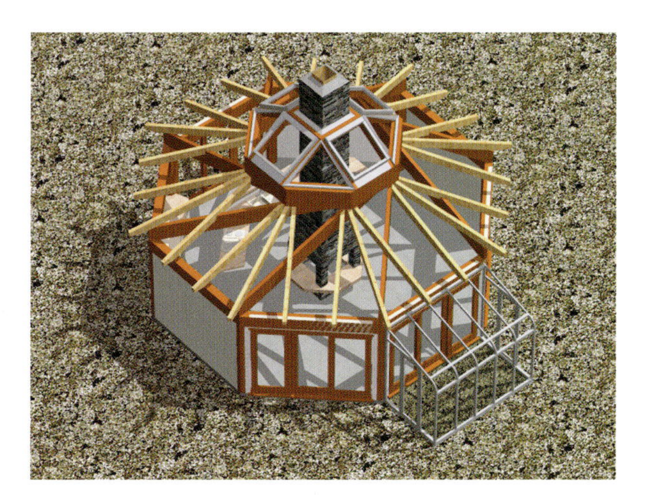

▲ **11.7** Three-dimensional view of the structure of the house. (Drawing: Fred Oesch.)

The construction of the building took approximately 12 months.

The construction costs were approximately US $165 000, plus approx. US $35 000 for the off-the-grid photovoltaics.

▲ 11.8 Fitting the roof light. (Photo: Fred Oesch.)

▲ 11.9 The finished skeleton. (Photo: Fred Oesch.)

▲ **11.10** The RF roof – internal view. (Photo: Fred Oesch.)

▲ **11.11** The roof light – internal view. (Photo: Fred Oesch.)

▲ 11.12 Internal view of the house. (Photo: Fred Oesch.)

▲ 11.13 The architect and builder in the house. (Photo: Mr Loony.)

Design process

The client wanted a house that would enable her to lead an autonomous lifestyle. Both the architect and the builder shared the client's vision, and with her encouragement and support they were able to suggest a bold design with some imaginative design solutions.

Fred Oesch, the architect for this project, explains:

> *'A reciprocal frame roof was chosen because of its simple affordable "kit-of-parts" modular nature. The free span structure allows for unlimited freedom in the placement or future relocation of interior walls. Furthermore, the exposed interior spiral structure is exciting and spiritually uplifting, rather like a chapel or place of meditation.'*

The architect, builder and client, by working together, have created a design that is special for its 'green' credentials, but is also one that inspires architecturally.

▲ **11.14** An RF fruit bowl designed by Fred Oesch. (Photo: Fred Oesch.)

Spey Valley Reciprocal Frame House

Background

In the Spey Valley, only a few miles from the town of Laggan in the Scottish Highlands, Roy Tilden Wright undertook the building of his own ideal home. He imagined the house standing on a rise in the valley overlooking the Spey beyond. He approached Out of Nowhere (OON), the design-build company founded by designer Graham Brown, after having been through design processes with several other architects and designers only to be disappointed by their lack of imagination, and lack of enthusiasm for a collaborative design process. At OON Roy was introduced to Hugh Adamson, at that stage a relatively young architecture graduate from Australia. It was in this partnership that Roy found the collaborative design approach he was looking for. The client opted for an hourly charge arrangement, not wanting the design and procurement to be held up by set time-frames.

▲ **11.15** Spey Valley house: North façade. (Computer-produced image provided by Hugh Adamson.)

▲ **11.16** South façade. (Computer-produced image provided by Hugh Adamson.)

Design process

Design began in November 2003 and the application for a building permit was submitted in September 2005. Planning was complicated by the fact that the land, though freehold, is within the National Park, and consequently the highly unusual structure had to pass through two layers of planning.

Roy chose a reciprocal frame as part of his dedication to doing something unique, and in coming to OON he knew that was what he could expect.

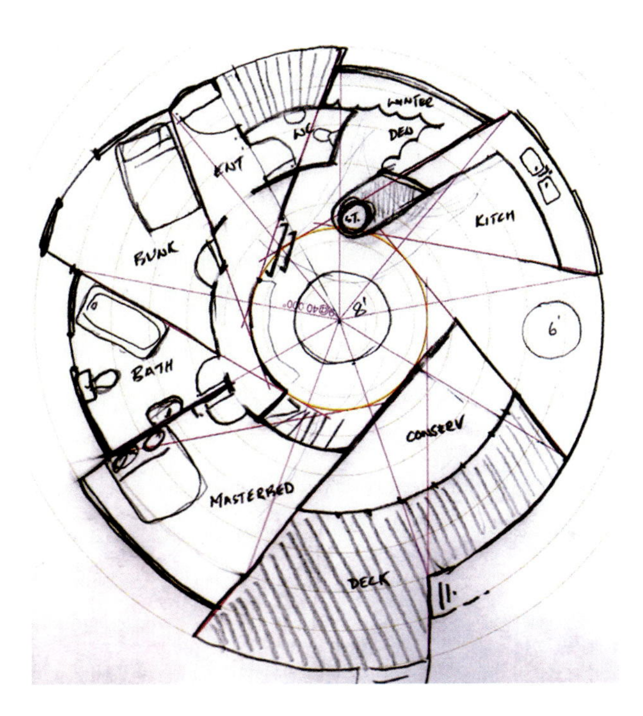

▲ **11.17** Plan – sketch. (Drawing: Hugh Adamson.)

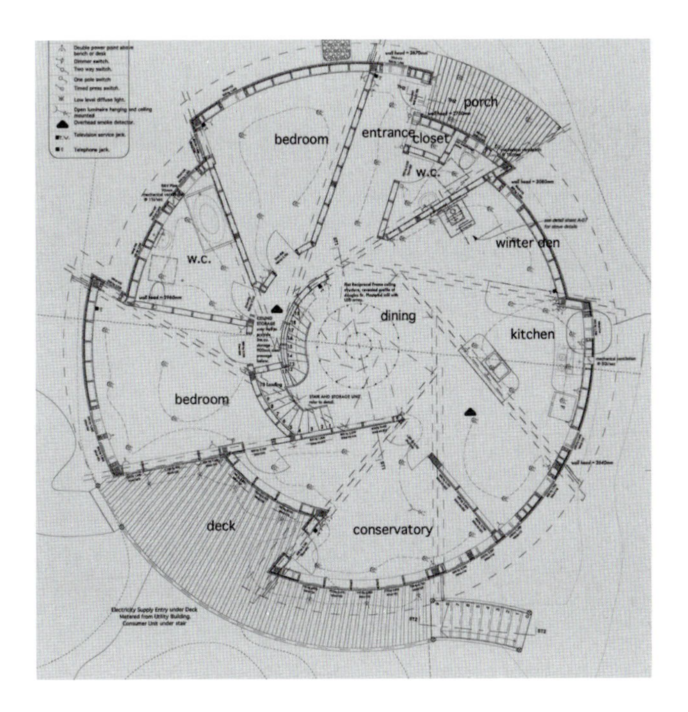

▲ **11.18** Finished plan. (Drawing: Hugh Adamson.)

The house design

There are several elements to this design that are, to the design team's knowledge, the first of their kind. The first is the flexibility employed with the floor plan and the consequent idiosyncratic pushing and pulling of the volumes from the centre of what would normally be a symmetrical plan design. So while sticking to the geometric division of a nine-sided reciprocal frame, a number arrived at both intuitively and practically, the distance of these 'sides' from the centre varied according to function.

The faceted roof used together with a curved wall was also untested and this, combined with various wall heights due to their distance from the centre (the further from the centre, the lower the wall), meant that nearly all wall heads and columns were unique. The construction strategy employed to achieve the exact stud height through space – a curved wall meeting a roof plane of compound pitch without the use of a head plate – was to build the roof oversize and prop it up, then plumb in the studs from the plan and bolt them off where they met a rafter. The excess roof was then cut to the desired eave overhang.

▲ **11.19** Perspective view. (Drawing: Hugh Adamson.)

Perhaps the greatest divergence from convention was the case of the beams themselves where, out of a desire for maximum slenderness, beam depths were arrived at by their individual loading instead of their highest common loading. Because of this, the beam cuts became very complicated, and it was only through the use of 3D modelling and 'Boolean extraction' that the beam cuts could be defined for machining.

The final, hitherto untested element of the design was placing a room at the top of the reciprocal frame, hanging a floor from the beam connections and running a stair through what is typically a much smaller atrium void. The floor to this upper chamber is itself a nine-sided flat reciprocal floor frame. Other than the concrete fins rising out of the hill as footings, all the other structural members are timber. The reciprocal frame floor beams are oak, the framing is construction grade pine and the roof beams are straight glued laminated timber, which are either 145×450, 145×495 or 165×540 mm in size.

Client/builder: Roy Tilden Wright

Architectural design: Hugh Adamson of Out of Nowhere

Structural design: Peter Murray of Leonard Murray Associates

Graham Brown and Scott Gamble of Out of Nowhere must also be mentioned for the many hours of assistance in working through how exactly to put the building together.

12 Postscript

There are many factors that will influence the design of a building: a synthesis of considerations related to the site, the historical context, the function of the building, the aesthetic appearance, building physics and other issues. The structural system will be only one of them. We judge the quality of a design on how harmoniously the synthesis of the multitude of influential factors has been achieved.

One can argue that for different projects and for different people the level of importance of the influential factors will vary. Regardless of the fact that there always will be a level of subjectivity in judging design, in most cases the masterpieces and the failures are easy to spot and agree upon.

In the author's opinion, this book presents some real architectural masterpieces, especially when looking at the work of Japanese designers Kazuhiro Ishii, Yoichi Kan and Yasufumi Kijima.

Although structure is only one of the multitude of, at times opposing, factors that influence building design, there are instances when the structure becomes part of the overall narrative, form and architectural expression. More importantly, when it forms and is part of the harmonious composition that we class as architecture, it is an influential factor that, to a lesser or greater degree, determines the level of success of a building design. And although the structure as such cannot determine the quality of a building design, if integrated appropriately it can influence it greatly.

Reciprocal frames are presented here as simply one more option that is available for building design. It is a system that offers great opportunities but also has its limitations. I hope that by introducing readers to the world of reciprocal frame architecture, it may inspire talented and skilled design teams to create new and imaginative buildings using reciprocal frames.

BIBLIOGRAPHY

Alexandar, C. (1979). *The Timeless Way of Building*. Oxford University Press.

Anon (1956). *Leonardo Da Vinci*, memorial edition based on the Leonardo Exposition held in Milan in 1939. Reynal & Company, New York.

Anon (1995). *The Structural Engineer*, Vol. 73, No. 4, February.

Ardalan, N. and Bakhtiar, L. (1973). *The Sense of Unity – The Sufi Tradition in Persian Architecture*. University of Chicago Press, Chicago.

Auboyer, N. (1967). *The Oriental World – India and South-East Asia*. Hamlyn, London.

Baker, H. B. (1989). *Design Strategies in Architecture*. Van Nostrand Reinhold.

Balmond, C. (2002). *Informal*. Prestel.

Baverel, O. (2000). *Nexorades: A Family of Interwoven Space Structures*. Ph.D. thesis, University of Surrey.

Bertin, V. (2001). Variations of lever beam structures. In *Proceedings of Conference on Growth and Form: The Engineering of Nature*. School of Architecture, University of Waterloo.

Bertin, V. (2002). Hebelstabwerke. In *ARCH + 159/160* (in German), May.

Bertin, V. and Lonnman, B. (2001). A study of form: mutually supported stick structures. In *Proceedings of Paradoxes of Progress*, 89th ACSA Annual Meeting, Baltimore, MD.

Bowie, T. (1959). *The Sketchbook of Villard de Honnecourt*. Indiana University Press, Bloomington.

Brown, G. (1987). *Out of Nowhere – New Structures to Live Within*. Unpublished.

Burry, M. and Popovic, O. (1997). *Proceedings of the International Colloquium: Structural Morphology – Towards the New Millennium*. Reciprocal Frame (RF) Parametric Studies, Nottingham, August.

Chilton, J. C. (1994). Polygonal living – some environment-friendly buildings with reciprocal frame roofs. In *Proceedings International Seminar on Structural Morphology in Architecture*, Stuttgart, Germany, pp. 21–29.

Chilton, J. C. (1995). History of timber structures, Lecture E1. In *STEP 2, Timber Engineering* (Blass, H. J. et al., eds), Vol. 2, pp. E1–E13. Centrum Hout, The Netherlands.

Chilton, J. C. and Choo, B. C. (1992). Reciprocal frame long span structures. In *Proceedings of the International Association for Shell and Spatial Structures*, Canadian Society of Civil Engineers, International Congress on Innovative Large Span Structures, Montreal, Canada, July, Vol. 2, pp. 100–109.

Chilton, J. C., Choo, B. C. and Coulliette, P. (1994). Retractable roofs using the 'reciprocal frame'. In *Proceedings from the IABSE*, Birmingham, UK, September, pp. 49–54.

Chilton, J. C., Choo, B. C. and Yu, J. (1994). Morphology of reciprocal frame 3-dimensional grillage structures. In *Spatial Lattice and Tension Structures, Proceedings IASS-ASCE* (Abel, J. F., Leonard, J. W. and Penalba, C. U., eds), Atlanta, GA, pp. 1065–1072.

Chilton, J. C., Choo, B. S. and Popovic, O. (1995). Reciprocal frames past, present and future. In *Proceedings of the International Conference in Lightweight Structures in Civil Engineering* (Obrebski, J. B., ed.), Warsaw, Poland, September, pp. 26–29. Magat-Magdalena Burska, Warsaw, Poland.

Chilton, J. C., Choo, B. S. and Popovic, O. (1995). Reciprocal frame 3-dimensional grillage structures. In *Proceedings of the International Conference in Lightweight Structures in Civil Engineering* (Obrebski, J. B., ed.), Warsaw, Poland, September, pp. 75–79. Magat-Magdalena Burska, Warsaw, Poland.

Crossley, F. H. (1951). *Timber Building in England, from Early Times to the End of the Seventeenth Century*. Batsford, London.

Detail (1994). *Puppet Theatre in Seiwa*, No. 3, pp. 322–325.

Emy, A. R. (1841). *Traite de L'art de la Charpenterie, Atlas*. Garilian-Geyry and V. Dalmont, Paris.

Evans, D. G. (1987). *The Structural Engineer*, Vol. 65A, No. 6, June.

Flores, C. (1982). *Gaudi, Jujol y el Modernismo Catalan*. Aguilar, Spain.

Gat, D. (1992). Mutually restrained, modular floating platforms. In *Proceedings of International Congress in Innovative Large Span Structures*, Montreal, Vol. 1, pp. 859–868.

Giurgola, R. (1979). *Louis I. Kahn*. Verlag fur Architectur, Artemis, Zurich.

Gombrich, E. H. (1979). The sense of order. *A Study of the Psychology of Decorative Art*. Phaidon Press, Oxford.

Hansen, J. (1971). *Architecture in Wood*. Faber & Faber, London.

Harison, R. (1991). *The Built, the Unbuilt and the Unbuildable, in Pursuit of Architectural Meaning*. Thames and Hudson.

Hartoonian, G. (1994). *Ontology of Construction*. Cambridge University Press.

Hewett, C. A. (1974). *English Cathedral Carpentry*. Wayland Publishers, London.

Inoue, M. (1985). *Space in Japanese Architecture*. John Weatherhill, New York.

Ishii, K. (1978). Sukiya concept. *GA Houses*, Vol. 4, pp. 249–304.

Ishii, K. (1990). *Sukiya Village and 51 Other Works, Space Design*. Kajima Institute Publishing, Japan.

Ishii, K. (1992/3). Seiwa Bunraku Puppet Theatre. *Japan Architect*, No. 2, pp. 44–51.

Ishii, K. (1994). Puppet Theatre in Seiwa, Japan. *Detail*, March, pp. 322–325.

Itoh, T. (1969). *The Elegant Japanese House, Traditional Sukiya Architecture*. Walker/Weatherhill, New York.

Itoh, T. (1972). *Traditional Domestic Architecture of Japan*, 2nd edition. Weatherhill/Heibonsha, Tokyo.

Itoh, T. (1974). *The Classic Tradition in Japanese Architecture*. Weatherhill/Tankosha, New York.

Jodidio, P. (2006). *Architecture in Japan*. Taschen.

Kirashiki, S. (1995). *Yasufumi Kijima 1972–1994* (in Japanese).

Kurokawa, K. (1991). *New Wave Japanese Architecture*, academy editions. Ernst & Son.

Laugier, M. A. (1977). *An Essay on Architecture*, first edited in 1755. Hennessay & Ingalls, Los Angeles.

Masuda, T. (1970). *Living Architecture: Japanese*. Macdonald & Co., London.

Maxwell, C. J. (1890). On reciprocal figures, frames, and diagrams of forces. In *The Scientific Papers of James Clerk Maxwell* (Niven, W. D., ed.). Cambridge University Press.

Meyhofer, D. (1994). *Contemporary Japanese Architects*. Benedikt Taschen.

Murray, P. (1986). *The Architecture of the Italian Renaissance*, 3rd edition. B.T. Batsford Ltd, London.

Natterer, J., Herzog, T. and Volz, M. (1991). *Holzbau Atlas Zwei*. Institut fur Internationale Architectur, Munich.

Nepilly, E. et al. (2004). *Tokyo Architecture and Design*. TeNeues.

Nishi, K. and Hozumi, K. (2004). *What is Japanese Architecture: A Survey of Traditional Japanese Architecture*. Kodansha, Japan.

Oliver, P. (1987). *Dwellings: The House Across the World*. Phaidon Press, Oxford.

Osborne, H. (1972). *Aesthetics*. Oxford University Press.

Pearson, D. (1994). *Earth to Spirit: In Search for Natural Architecture*. Gaia Books, London.

Popovic, O. (1996). *Reciprocal Frame Structures*. Ph.D. thesis, University of Nottingham.

Popovic Larsen, O. and Tyas, A. (2003). *Conceptual Structural Design: Bridging the Gap between Engineers and Architects*. Thomas Telford Books.

Popovic, O., Chilton, J. C. and Choo, B. S. (1996). Rapid construction of modular buildings using the 'reciprocal frame'. In *Proceedings of the International Conference on Mobile and Rapidly Assembled Structures* (Brebbia, C. A., ed.), MARAS, Seville, Spain, June, pp. 73–82. Computational Mechanics Publications.

Popovic, O., Chilton, J. C. and Choo, B. S. (1996). Sustainable roundwood reciprocal frame structures. In *Proceedings of the International Conference Detail Design in Architecture* (Emmitt, S., ed.), Northampton, September.

Popovic, O, Chilton, J. C. and Choo, B. S. (1998). The variety of reciprocal frame (RF) morphologies developed for a medium span assembly building. *Journal of the International Association for Shell and Spatial Structures*, Vol. 39, No. 1, pp. 29–35.

Popovic Larsen, O., Davison, J. B. and Tyas, A. (2002). Architecture and engineering – a joint education at Sheffield University. *International Journal of Space Structures – Special Edition on Teaching of Space Structures* (Chilton, J., ed.), July, pp. 205–213.

Process – Architecture (1981). Japan: Climate, space and concept. No. 25.

Richter, J. P. (1977). *The Literary Works of Leonardo da Vinci*. Phaidon, Oxford.

Rizzuto, J. P. (2005). *The Structural Behavior of Mutually Supported Elements in Space Structures*. Ph.D. thesis, Coventry University.

Rizzuto, J. P. (2006). Notched mutually supported element (MSE) circuits in space structures. In *Proceedings of the IASS Symposium New Olympics – New Shell and Spatial Structures*, Beijing, pp. 180–182.

Rizzuto, J. P., Saidani, M. and Chilton, J. C. (2000). The self-supporting multi-reciprocal grid (MRG) system using notched elements. *Journal of the International Association for Shell and Spatial Structures,* Vol. 41, No. 2, pp. 125–131.

Rowan, J. (1968). Editorial. *Progressive Architecture*, November, p. 93.

Rowland, K. (1964). *Looking and Seeing 1 – Pattern and Shape*. Ginn & Co., Aylesbury.

Saidani, M. and Baverel, O. (1998). Investigation into a new type of multi-reciprocal grid. *International Journal of Space Structures*, Vol. 13, No. 4, pp. 215–218.

Saidani, M. and Baverel, O. (1998). Retractable multi-reciprocal grid structure. *Journal of the International Association for Shell and Spatial Structures*, Vol. 39, No. 2, pp. 141–146.

Schmertz, M. F. (1994). Structural inventor. *Architecture*, June, pp. 101–107.

Scully, V. Jr (1962). *Louis I. Kahn*. George Braziller, New York.

Serlio, S. (1970). *First Book of Architecture by Sebastiano Serlio*, first published 1619. Benjamin Bloom, New York.

Space Design 9404 (1994). *All Works of Yasufumi Kijima*, pp. 110–115.

Sumiyoshi, T. and Matsui, G. (1990). *Wood Joints in Classical Japanese Architecture*. Kajima Institute Publishing Co., Japan.

Suzuki, H., Banham, R. and Kobayashi, K. (1985). *Contemporary Architecture of Japan 1958–1984*. The Architectural Press, London.

Tredgold, T. (1890). *Elementary Principles of Carpentry*. E. and F. N. Spon, The Strand, London.

Wallis, J. (1972). *Opera Matematica* (in Latin), first published 1695. Georg Olms Verlag Hildesheim, New York.

Wrench, T. (2001). *Building a Low Impact Roundhouse*. Permanent Publications, UK.

www.digilanderlibero.it
www.findhorn.org
www.fredoesch.com
www.kimwilliamsbooks.com
www.rinusroelofs.nl
www.thatroundhouse.info
www.woodlandburialparks.co.uk

Index

[Italic numerals refer to illustrations]